OXFORD MEDICAL ENGINEERING SERIES

Editors: B. McA. Sayers, J. I. E. Hoffman

OXFORD MEDICAL ENGINEERING SERIES

1. The theoretical basis of electrocardiology
   Edited by C. V. Nelson and D. B. Geselowitz
2. Auditory investigation: the scientific and technological basis
   Edited by H. A. Beagley
3. The study of heart-rate variability
   Edited by R. I. Kitney and O. Rompelman
4. Human body dynamics
   Edited by D. N. Ghista
5. Biomechanical measurement in orthopaedic practice
   Edited by M. W. Whittle and J. D. Harris

Biomechanical measurement in orthopaedic practice

EDITED BY

MICHAEL WHITTLE

Consultant Clinical Physiologist, Oxford Orthopaedic Engineering Centre

and

DEREK HARRIS

Director, Oxford Orthopaedic Engineering Centre

CLARENDON PRESS · OXFORD
1985

Oxford University Press, Walton Street, Oxford OX2 6DP

Oxford New York Toronto
Delhi Bombay Calcutta Madras Karachi
Kuala Lumpur Singapore Hong Kong Tokyo
Nairobi Dar es Salaam Cape Town
Melbourne Auckland

and associated companies in
Beirut Berlin Ibadan Mexico City Nicosia

Oxford is a trade mark of Oxford University Press

Published in the United States
by Oxford University Press, New York

British Library Cataloguing in Publication Data

Biomechanical measurement in orthopaedic practice.
—(Oxford medical engineering series; 5)
1. Orthopedia 2. Biomechanics
I. Whittle, Michael II. Harris, Derek
617'.3 RD732
ISBN 0-19-857610-2

Library of Congress Cataloging in Publication Data
Main entry under title:

Biomechanical measurement in orthopaedic practice.

(Oxford medical engineering series; 5)
Includes bibliographies and index.
1. Orthopedia. 2. Human mechanics. I. Whittle, Michael. II. Harris, J. Derek. III. Series.
[DNLM: 1. Biomechanics. 2. Orthopedics. WE 103 B614]
RD732.B56 1985 617'.3 84-23308
ISBN 0-19-857610-2

Set by Joshua Associates, Oxford
Printed in Great Britain by
St Edmundsbury Press,
Bury St Edmunds, Suffolk

Preface

Biomechanical studies in orthopaedics present an urgent challenge to the clinician who wishes to have a better understanding of the mechanical forces which are influencing the progress and the treatment of his patient.

It has been demonstrated beyond doubt that the human eye cannot objectively record the subtle changes which in some cases are all that define improvement or deterioration of the disease, or even the range and variation of what can be described as 'normal'.

But the biomechanical laboratory is often situated remotely from the clinic and the terminology is strange, even giving different meanings to clinical words. Test results may go against established opinion.

Clinical practice however is becoming accustomed to accelerating change. CT scanning and nuclear magnetic resonance are examples of techniques becoming accepted into routine use, and leaders in orthopaedic practice accept the need to learn the new forms. A problem however exists for the clinician when he is inadequately prepared in biomechanics by his training. But the high proportion of research papers on this subject by clinicians show that this can be overcome. The other main difficulty is that results of patient analyses can seldom be distilled into single figures or simple statements.

This book is compiled for clinicians who are keen to obtain a better understanding of the current status of biomechanical measurement in orthopaedic practice. It is also for research workers wishing to know more about techniques they may be able to incorporate in their work. However, its main purpose is to help in building the bridge between orthopaedics and engineering.

The opportunity for this book arose from a conference held in the Nuffield Orthopaedic Centre in Oxford in 1983 and the editors are most grateful to all those who contributed to this conference.

Oxford M.W.W.
October 1983 J.D.H.

Contents

SECTION 4: UPPER AND LOWER LIMBS

Contributors

J. Albert, Department of Orthopaedics, Norfolk and Norwich Hospital, Norwich, England.

D. L. Bader, Oxford Orthopaedic Engineering Centre, Nuffield Orthopaedic Centre, Oxford, England

G. Benoni, Department of Orthopaedic Surgery, Royal National Orthopaedic Hospital, London, England.

G. Bentley, Department of Orthopaedic Surgery, Royal National Orthopaedic Hospital, London, England.

R. Bourgois, Department of Orthopaedics and Traumatology, Hôpital Erasme, University of Brussels, Belgium.

T. R. M. Brown, Bioengineering Unit, University of Strathclyde, Glasgow, Scotland.

F. Burny, Department of Orthopaedics and Traumatology, Hôpital Erasme, University of Brussels, Belgium.

S. J. Burrough, Horton General Hospital, Banbury, England.

P. G. Carter, Department of Anatomy, King's College, London, England.

P. R. Cavanagh, Biomechanics Laboratory, The Pennsylvania State University, Pennsylvania, USA.

L. C. Cobb, Department of Orthopaedics and Rehabilitation, University of Vermont College of Medicine, Burlington, USA.

G. V. B. Cochran, Orthopaedic Engineering and Research Center, Helen Hayes Hospital, West Haverstraw, New York, USA.

M. Cornelissen, ICOBI Biomechanics and Biomaterials Section, University of Leuven, Belgium.

R. Dee, Department of Orthopaedics, Health Sciences Center, State University of New York at Stony Brook, Stony Brook, USA.

D. M. Denison, The Lung Function Unit, Brompton Hospital, London, England.

G. J. Docter, Department of Anatomy, Free University, Amsterdam, The Netherlands.

M. Donkerwolcke, Department of Orthopaedics and Traumatology, Hôspital Erasme, University of Brussels, Belgium.

G. S. E. Dowd, Royal National Orthopaedic Hospital, London, England.

T. Duckworth, Department of Orthopaedics, Royal Hallamshire Hospital, Sheffield, England.

J. Ensink, Department of Biomechanics, Free University, Amsterdam, The Netherlands.

J. H. Evans, Bioengineering Unit, University of Stratchlyde, Glasgow, Scotland.

M. Evans, Oxford Orthopaedic Engineering Centre, Nuffield Orthopaedic Centre, Oxford, England.

J. C. T. Fairbank, Department of Orthopaedics, Norfolk and Norwich Hospital, Norwich, England.

M. W. Ferguson-Pell, Orthopaedic Engineering and Research Center, Helen Hayes Hospital, West Haverstraw, New York, USA.

J. W. Frymoyer, Department of Orthopaedics and Rehabilitation, University of Vermont College of Medicine, Burlington, USA.

A. R. Gourlay, The Lung Function Unit, Brompton Hospital, London, England.

J. Gwillim, Oxford Orthopaedic Engineering Centre, Nuffield Orthopaedic Centre, Oxford, England.

D. L. Hamblen, Department of Orthopaedics, University of Glasgow, Scotland.

L. I. Hansson, Department of Diagnostic Radiology, University of Lund, Lund, Sweden.

Derek Harris, Oxford Orthopaedic Engineering Centre, Nuffield Orthopaedic Centre, Oxford, England.

I. Haslock, Bioengineering Group, Department of Engineering, University of Durham, Durham, England.

E. M. Hennig, Biomechanics Laboratory, The Pennsylvania State University, Pennsylvania, USA.

C. L. Hubley, School of Physiotherapy, Dalhousie University, Halifax, Nova Scotia, Canada.

L. Hurst, Department of Orthopaedics, Health Sciences Center, State University of New York at Stony Brook, Stony Brook, USA.

R. J. Jefferson, Oxford Orthopaedic Engineering Centre, Nuffield Orthopaedic Centre, Oxford, England.

F. Johnson, Bioengineering Research Unit, University of Nottingham, Nottingham, England.

A. R. Jones, Bioengineering Group, Department of Engineering, University of Durham, Durham, England.

I. G. Kelly, Department of Orthopaedics, University of Glasgow, Scotland.

J. Kenwright, Nuffield Orthopaedic Centre, Oxford, England.

M. A. Lafortune, Biomechanics Laboratory, The Pennsylvania State University, Pennsylvania, USA.

A. Lindstrand, Department of Diagnostic Radiology, University of Lund, Lund, Sweden.

K. Lindström, Department of Biomedical Engineering, University Hospital, Malmö, Sweden.

K. Linge, Royal Liverpool Hospital, Liverpool, England.

M. Lord, Bioengineering Centre, University College, London, England.

L. L. Lowery, Bioengineering Research Unit, University of Nottingham, Nottingham, England.

M. T. Manley, Department of Orthopaedics, Health Sciences Center, State University of New York at Stony Brook, Stony Brook, USA.

L. Mauritzson, Department of Biomedical Engineering, University Hospital, Malmö, Sweden.

K. Mills, National Hospital for Nervous Diseases, London, England.

B. Mjöberg, Department of Orthopaedics, University of Lund, Lund, Sweden.

M. S. Moreland, Department of Orthopaedics and Rehabilitation, University of Vermont College of Medicine, Burlington, USA.

M. D. L. Morgan, The Lung Function Unit, Brompton Hospital, London, England.

A. Moulton, Department of Orthopaedic Surgery, Mansfield and District General and Harlow Wood Orthopaedic Hospital, Mansfield, England.

M. Mulier, ICOBI Biomechanics and Biomaterials Section, University of Leuven, Belgium.

T. P. Newson, Oxford Orthopaedic Engineering Centre, Nuffield Orthopaedic Centre, Oxford, England.

S. H. Norris, Department of Orthopaedics, Northern General Hospital, Sheffield, England.

J. P. O. O'Brien, ORLAU, Robert Jones and Agnes Hunt Orthopaedic Hospital, Oswestry, England.

V. Palmieri, Orthopaedic Engineering and Research Center, Helen Hayes Hospital, West Haverstraw, New York, USA.

J. P. Paul, Bioengineering Unit, University of Strathcylde, Glasgow, Scotland.

A. J. Peacock, The Lung Function Unit, Brompton Hospital, London, England.

M. J. Pearcy, Oxford Orthopaedic Engineering Centre, Nuffield Orthopaedic Centre, Oxford, England.

H. Phillips, Department of Orthopaedics, Norfolk and Norwich Hospital, Norwich, England.

M. H. Pope, Department of Orthopaedics and Rehabilitation, University of Vermont College of Medicine, Burlington, USA.

I. Portek, Oxford Orthopaedic Engineering Centre, Nuffield Orthopaedic Centre, Oxford, England.

M. L. Porter, Department of Orthopaedics, Manchester Royal Infirmary, Manchester, England.

E. S. Powell, Department of Medical Physics, Royal Hallamshire Hospital, Sheffield, England.

D. J. Pratt, Orthotics and Disability Research Centre, Derbyshire Royal Infirmary, Derby, England.

W. K. Purves, Department of Medical Physics, Manchester Royal Infirmary, Manchester, England.

P. B. Pynsent, Department of Orthopaedics, Norfolk and Norwich Hospital, Norwich, England.

N. P. Reddy, Institute for Biomedical Engineering Research, University of Akron, Akron, Ohio, USA.

M. M. Rodgers, Biomechanics Laboratory, The Pennsylvania State University, Pennsylvania, USA.

G. K. Rose, Gait Assessment Laboratory, Orthotic Research and Locomotor Assessment Unit, Robert Jones and Agnes Hunt Orthopaedic Hospital, Oswestry, England.

D. Rossmere, Department of Orthopaedics, Health Sciences Center, State University of New York at Stony Brook, Stony Brook, USA.

D. I. Rowley, Department of

Orthopaedics, Royal Hallamshire Hospital, Sheffield, England.

L. Ryd, Department of Orthopaedics, University of Lund, Lund, Sweden.

D. J. Sanderson, Biomechanics Laboratory, The Pennsylvania State University, Pennsylvania, USA.

O. Saric, Department of Orthopaedics and Traumatology, Hôpital Erasme, University of Brussels, Belgium.

G. Selvick, Department of Anatomy, University of Lund, Lund, Sweden.

D. Shakespeare, Nuffield Orthopaedic Centre, Oxford, England.

J. Shaw Dunn, Bioengineering Unit, University of Strathclyde, Glasgow, Scotland.

J. E. Shepherd, Department of Human Sciences, University of Loughborough, Loughborough, England.

K. Sherman, Nuffield Orthopaedic Centre, Oxford, England.

D. M. Smith, Bioengineering Centre, University College, London, England.

R. W. Soames, Department of Anatomy, King's College, London, England.

S. F. C. Stewart, Orthopaedic Engineering and Research Center, Helen Hayes Hospital, West Haverstraw, New York, USA.

I. Stockley, Department of Orthopaedics, Manchester Royal Infirmary, Manchester, England.

I. A. F. Stokes, Department of Orthopaedics and Rehabilitation, University of Vermont College of Medicine, Burlington, USA.

M. Stokes, Nuffield Orthopaedic Centre, Oxford, England.

K. E. Tanner, Oxford Orthopaedic Engineering Centre, Nuffield Orthopaedic Centre, Oxford, England.

K. M. Tesh, Bioengineering Unit, University of Strathclyde, Glasgow, Scotland.

J. A. Towle, Department of Anatomy, King's College, London, England.

A. R. Turner-Smith, Oxford Orthopaedic Engineering Centre, Nuffield Orthopaedic Centre, Oxford, England.

A. Unsworth, Bioengineering Group, Department of Engineering, University of Durham, Durham, England.

S. S. Upadhyay, Department of Orthopaedic Surgery, Mansfield and District General Hospital and Harlow Wood Orthopaedic Hospital, Mansfield, England.

G. Van der Perre, ICOBI Biomechanics and Biomaterials Section, University of Leuven, Belgium.

J. C. Wall, School of Physiotherapy, Dalhousie University, Halifax, Nova Scotia, Canada.

Michael Whittle, Oxford Orthopaedic Engineering Centre, Nuffield Orthopaedic Centre, Oxford, England.

S. Willner, Department of Orthopaedic Surgery, University Hospital, Malmö, Sweden.

B. P. Wordsworth, Department of Rheumatology, Nuffield Orthopaedic Centre, Oxford, England.

A. Young, Nuffield Orthopaedic Centre, Oxford, England.

Section 1:
Fracture Management

1 Biomechanical Measurement of Fracture Repair

J. KENWRIGHT

INTRODUCTION

Fractures are still one of the commonest reasons for spending prolonged periods of time away from work, and this morbidity occurs particularly following fractures of the long bones. The latest Hospital In-patient Enquiry in England and Wales showed an incidence of tibial fractures of 13 590 per year. Overall non-union rates for this group of fractures have usually been assessed as between 5 and 10 per cent, but delay in bone healing in excess of 20 weeks from the time of injury is present in over 25 per cent. In centres having to treat many road-traffic accidents, such fractures are often open with severe wounds, and in this group over 40 per cent of fractures may take over six months to heal (Rosenthal *et al.* 1977).

There is no evidence that preventive measures are decreasing the incidence of these violent fractures in the young and working population; nor, in the short term, is there likely to be a reduction in the numerous fractures of the lower-limb bones occurring in the elderly suffering from osteoporosis.

Little is known about the biomechanical events occurring during the healing of fractures and even the assessment of one end-point of healing—bone union—is a crude science. The diagnosis of the point of clinical union of a fracture is made usually by manual assessment of stability, followed by radiological examination. Two of the main functions of a long bone are support of weight and locomotion. It does seem important, therefore, that healing of a fracture should be assessed chiefly in terms of increasing mechanical strength. Therefore, the most relevant of the tests used in clinical practice is probably the manual test of stability. However, this is subjective and very inexact.

Clinical methods combined with radiological examination, however, are satisfactory for defining the end-point of fracture union in approximately 90 per cent of patients, though with an accuracy of ± 3 weeks for an average long-bone fracture.

There are, however, three problem areas where the normal clinical methods are inadequate:

1. The difficult diagnostic problem.

2. For comparing treatment methods where there may be subtle differences in healing rates.

3. For sequential investigation of the phases of fracture healing both in clinical and experimental research on: (i) the intermediate mechanical stages of fracture healing; and (ii) the optimal mechanical environment for rapid fracture healing.

Examples are now given of the difficulties that can be encountered in these problem areas. This will be followed by a review of current methods of assessment and how they have been used in clinical practice and in research.

THE DIFFICULT DIAGNOSTIC PROBLEM

1. The accurate definition of the mechanical integrity of the healed fracture may be very important in patients returning to strenuous work or professional sport. Patients in these categories who have sustained tibial or femoral diaphyseal fractures may be at risk of refracture; the normal clinical and radiological methods of assessment are very inaccurate methods of defining the mechanical integrity of the healed bone. In our present state of knowledge these patients may either be placed at risk of refracture, or be prevented from taking part in normal activities for many months longer than is necessary.

2. In certain long-bone shaft fractures the healing process is modified by the method of treatment so that clinical assessment of mechanical integrity is impossible, and the interpretation of radiographs may be difficult. Diaphyseal fractures treated by 'rigid' internal plate fixation always demonstrate this problem, since the fracture cannot be manually tested mechanically and external callus formation is not seen on radiographs. Methods are needed to assess the mechanical integrity of facture healing in such circumstances, or an unreliable and unsafe rehabilitation programme may be prescribed.

3. Very severe tibial fractures treated using external skeletal fixation may also cause difficulty in defining when healing is mechanically sound enough to remove fixation and allow free weight bearing. In these patients there is often delayed healing and suppressed callus formation owing to local damage to the bone ends, the soft tissues, and their blood supply. There is also osteoporosis and the combination of these factors makes diagnosis of fracture union and fracture strength difficult.

4. Some fractures, such as those through the waist of the scaphoid bone in the wrist region, cause major difficulty in assessing bone healing. The bone is small and radiological assessment of healing very inexact.

5. Clinical and radiological assessment of non-union can be difficult. Because of inadequate diagnostic facilities, it may not be possible to confirm the presence of established non-union. Months may pass before the diagnosis is clear and the necessary surgical action taken. Assessment of the mechanical progress of healing fractures is also important in defining early those fractures, particularly

of long bones, where there is lack of progress, and in whom early surgical intervention with bone grafting would accelerate the healing process dramatically.

6. Finally, it has been shown that there is a 2 per cent risk of refracture after removal of implanted plates from 'healed' diaphyseal tibial or femoral fractures. Advances in assessment of the mechanical integrity of the healed fracture and adjacent bone before and after removal of implants of this type could be very important. It is a devastating event for a person to sustain a refracture after perhaps a long and difficult period of treatment.

COMPARING TREATMENT REGIMENS

It is essential to have an objective, accurate, and repeatable method of recording biomechanical end-points for fracture healing, if different treatment methods are to be compared. The controversy surrounding the use of electrical stimulation for enhancement of fracture healing would be lessened considerably if the end-points of non-union, delayed union, and subsequent bone healing could be defined in mechanical terms.

INVESTIGATIONS OF STAGES OF FRACTURE HEALING IN CLINICAL AND EXPERIMENTAL RESEARCH

There is a considerable literature describing research into the histological, microvascular, and biochemical events occurring during fracture healing, but very little information available about the sequence of biomechanical events. Further investigation is needed of: (i) the biomechanical changes occurring during healing; and (ii) the influence of different mechanical environments upon the stages of the fracture-healing process. Perhaps the mechanical conditions could be modified to obtain the optimal state for each stage of fracture healing. For example, in clinical practice perhaps a rigidly applied plate may be the best treatment immediately after fracture but at a later stage a different and more favourable mechanical environment might be appropriate.

METHODS OF ASSESSING FRACTURE HEALING AVAILABLE FOR USE IN CLINICAL PRACTICE

The following mechanical testing methods have been used in the assessment of fracture healing in patients.

1. Manual examination.
2. Radiological examination.
3. Direct deflection under load of attached but removable frame (Jernberger 1970).
4. Three-dimensional movement transducer using photodetectors (Tanner 1985).

5. Ultrasound (Gerlanc *et al.* 1975).
6. Resonant vibration (Cornelissen *et al.* 1985).
7. Stress wave propogation (Lewis 1975).
8. Strain-gauge monitoring of external skeletal fixator column (Burny 1979; Evans *et al.* 1985).
9. Scintigraphy (Auchincloss and Watt 1982).

It should be said that not many of the methods listed above have yet been used regularly as clinical tools for the non-invasive assessment of fracture healing.

The ideal method would be non-invasive, rapidly performed, and give an immediate result which would represent in a repeatable way the biomechanical qualities of the healing fracture at that time. The biomechanical profile should allow accurate prediction of the safe time of removal of fracture immobilization or predict the occurrence of non-union.

Both the vibration analysis technique described by Cornelissen *et al.* (1985) and the strain-gauge monitoring of the column in patients treated by external skeletal fixation have shown great potential in clinical practice. In our unit we have used this latter method and found it useful, especially for assessing the time of removal of external skeletal fixation in difficult cases with atrophic tibial fracture healing, or in those fractures with major bone loss treated using free-vascularized rib-bone grafts, when bone union must occur at multiple sites before the fixation frame can be removed. There have been strong clinical correlations (Evans *et al.* 1985).

EXPERIMENTAL METHODS AVAILABLE FOR ASSESSMENT OF FRACTURE HEALING

Many investigators have had to turn to experimental models to investigate fracture healing in order to achieve (i) a standard fracture; (ii) a standard protocol of treatment; and (iii) segmental examination of healing. However, in many of these studies it has not been possible to define and maintain the conditions in a uniform state with strict adherence to the protocol. It is difficult to standardize any fracture and certainly no experimental osteotomy resembles a motorcycle injury. Extrapolation to man from results in animals may also be misleading.

Despite these problems, most of our present knowledge concerning the biomechanics of fracture healing has been obtained from these types of investigation. In the experimental situation fracture healing can be studied biomechanically at intervals and the results correlated with radiographs, biochemistry, and histology. The mechanical environment of the healing fracture can be controlled and these same sequential observations made in groups of 'identical' fractures.

Sarmiento *et al.* (1977) demonstrated that stiffness for healing fractures in rats varied at different phases during consolidation of the fracture according to the degree of rigidity of fracture fixation imposed. White *et al.* (1977) applied regimens of intermittent loading to healing fractures and though they noted no consistent influence of this type of loading on fracture healing they did record distinct biomechanical stages of fracture healing (Stages 1-4) as assessed by torsional strength.

There is evidence from experimental studies on intact bone in sheep and birds that intermittent deformation applied cyclically enhances osteogenesis. The most important influences appear to be strain rate and strain magnitude (Goodship *et al.* 1979). In the study described here we have investigated the influence of altering the mechanical environment in a defined and controlled way during fracture healing (Kenwright *et al.* 1983). A fracture was simulated in sheep by performing a transverse osteotomy leaving a 3 mm gap. In one group of six sheep the osteotomy was fixed rigidly, using external skeletal fixation, and the sheep walked unrestricted. In the second group of six sheep the conditions at and after operation were the same as the first group, except that a controlled intermittent mechanical stimulus, with the qualities known to stimulate osteogenesis in intact bone, was applied axially for a very short period each day. The intermittent cyclical load was of constant force, and less than the stress limit of the bone-screw interface. A frequency of 0.5 Hz was used which is approximately that of physiological walking. An initial strain rate was applied of 30×10^3 μ/s which is above physiological levels, but had been found to be maximally osteogenic in intact bone. Five hundred cycles per day were applied, over 17 minutes, this also having been found to be highly osteogenic in intact bone. A 30 per cent strain rate was employed.

Fracture stiffness was measured every two weeks using strain gauge monitoring of the external fixation column (Fig. 1.1). There was no significant difference between the rigid fixation and the stimulated ('sliding clamp') group until 8-10 weeks from osteotomy, at which stage the stimulated group showed a statistically significant ($p < 0.05$) increase in stiffness. Similarly (Fig. 1.2), post-mortem torsional stiffness of the fracture expressed as a percentage of that of the intact tibia showed a statistically significant increase for the stimulated group ($p < 0.01$). Radiographs at two weeks showed callus formation in all instances in the stimulated group and none in the rigid group. It did appear, therefore, that this change in the mechanical environment, acting for a very short period each day, led to an enhanced rate of formation of early callus and to an increased torsional rigidity at 12 weeks from fracture.

CONCLUSIONS

In clinical practice there is a significant number of fractures for which the standard methods of clinical physical and radiological examination are inadequate. More sensitive and reliable investigations are needed if different treatment methods are to be assessed and compared objectively, and if a better understanding of the intermediate stages of fracture repair is to be achieved.

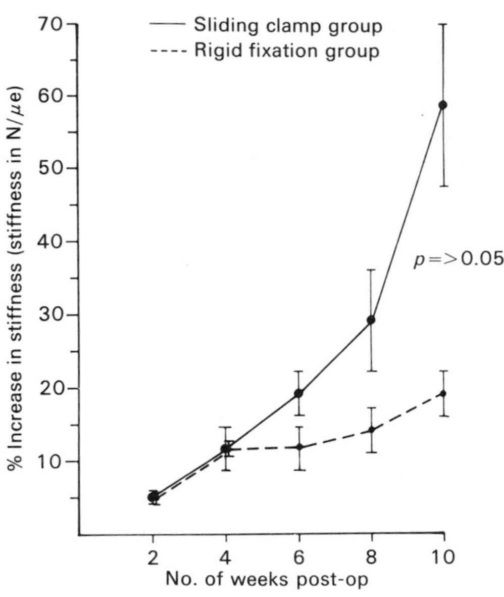

Fig. 1.1. The mean *in vivo* increase in fracture stiffness seen over 10 weeks when strain gauge monitoring an external skeletal fixation column.

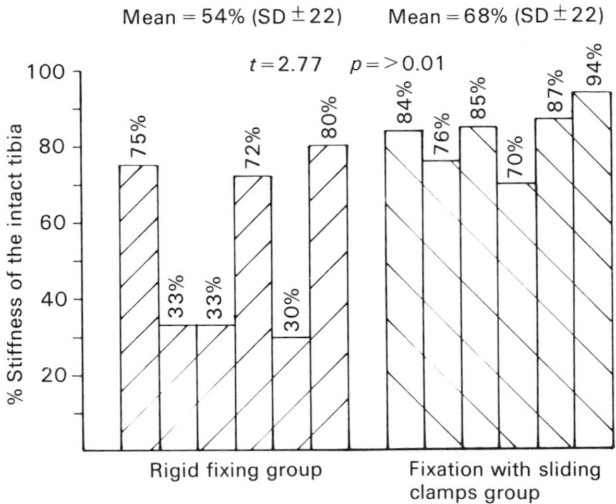

Fig. 1.2. The torsional stiffness of the tibia 12 weeks after operation for the rigidly fixed and intermittently loaded groups.

REFERENCES

Auchincloss, J. M. and Watt, I. (1982). Scintigraphy in the evaluation of potential fracture healing: a clinical study of tibial fractures. *Br. J. Radiol.* **55**, 707–13.

Burny, F. L. (1979). Strain gauge measurement of fracture healing. In *External fixation—the current state of the art* (ed. A. F. Brooker and C. C. Edwards) pp. 371–82. Williams & Wilkins, Baltimore.

Cornelissen, G., Van der Perre, G., and Burny, F. (1985). MADAMS—percussion technique: a non-invasive vibration analysis technique to monitor fracture healing. In *Biomechanical measurement in orthopaedic practice* (ed. M. W. Whittle and J. D. Harris) pp. 14–28. Oxford University Press.

Evans, M., Gwillim, J., Harris, J. D., and Tanner, K. E. (1985). Fracture monitoring. In *Biomechanical measurement in orthopaedic practice* (ed. M. W. Whittle and J. D. Harris) pp. 29–35. Oxford University Press.

Gerlanc, M., Haddad, D., Hyatt, G., Langloh, J., and St. Hilaire, P. (1975). Ultrasonic study of normal and fractured bones. *Clin. Orthopaed.* **111**, 175–80.

Goodship, A. E., Lanyon, L. E., and McFie, H. (1979). Functional adaptation of bone to increased stress. *J. Bone Jt Surg.* **61A**, 539–46.

Jernberger, A. (1970). Measurement of stability of tibial fractures. *Acta Orthopaed. Scand.* Suppl. 135.

Kenwright, J., Evans, M., Goodship, A. E., and O'Connor, J. (1983). The application of controlled intermittent cyclical loading in the acceleration of fracture healing. *J. Bone Jt Surg.* **65B**, 364.

Lewis, J. L. (1975). A dynamic model of a healing fractured long bone. *J. Biomechan.* **8**, 17–25.

Rosenthal, R. E., Malphail, J. A., and Ortiz, J. E. (1977). Non-union in open tibial fractures. *J. Bone Jt Surg.* **59A**, 244–8.

Sarmiento, A., Schaeffer, J. F., Beckerman, L., Latta, L. L., and Enis, J. E. (1977). Fracture healing in rat femora affected by functional weight bearing. *J. Bone Jt Surg.* **59A**, 369–75.

Tanner, K. E. (1985). The *in vivo* measurement of fracture movement. In *Biomechanical measurement in orthopaedic practice* (ed. M. W. Whittle and J. D. Harris) pp. 44–8. Oxford University Press.

White, A. A. III, Panjabi, M. M., and Southwick, W. O. (1977). Effects of compression and cyclical loading on fracture healing. *J. Biomechan.* **10**, 233–9.

2 Evaluation of Ultrasound Scanning in Healing Femoral Fractures

S. S. UPADHYAY AND A. MOULTON

INTRODUCTION

Ultrasound scanning is a harmless and simple non-invasive technique and facilities are available in almost all district hospitals in the United Kingdom. Its ability to provide cross-sectional views and visualization of cartilage along with bone gives it potential in diagnostic application in studying skeletal morphology, especially measuring the rotation or torsional variations of long bones and it can be repeated as many times as required without risk of exposure to radiation. Most ultrasonic examinations are currently performed using the echo-ranging or B-mode technique. Short waves of ultrasound are emitted by the transducer into the body, and the scanning image is based on the reflection of ultrasonic waves which occur at the boundaries between the different tissues within the body. A fraction of incident ultrasonic energy is reflected back if there is a change in the characteristic impedence as occurs at such a boundary. The energy which is not reflected travels beyond the boundary and may be reflected at deeper boundaries. The maximum penetration is limited by attenuation of the ultrasound in the tissue. Ultrasound is almost completely reflected from the surface of hard cortical bone. This prevents the examination inside or through such bone, thus one can examine only the outer surface of the cortical bone.

A direct method of measuring femoral anteversion using ultrasound was pioneered by Moulton and Upadhyay (1982). In this study, 30 dried femora were measured for femoral anteversion using a jig which would accept a femur and the angle of anteversion could be read directly on two protractors in line—one orientated proximally and the other distally. These femora were then re-measured for anteversion using ultrasound technique—the femur being submerged in a water-bath and scanned at two levels: one distally through the condylar region, and other proximally through the trochanteric region and the angle of anteversion was measured. The results gave the same readings except in one case where there was an unexplained discrepancy of 1°. Since then, the technique has been used in volunteers and patients with both orthopaedic conditions and fractures.

METHOD

Volunteers and patients

In order to establish the normal range of anteversion and variable difference between right and left hips, 30 adult volunteers were measured who had no obvious gait abnormality and who had no history or trauma or bone and joint disease. We also measured the angles of femoral anteversion in 25 patients who had had a unilateral fracture of the femoral shaft (average post-injury time of 6.3 years) in order to assess the range of persistent rotatory deformity which was present on accepted fracture management and a gross rotation deformity was not present clinically in most cases. The technique was used in 10 fresh femoral fractures whilst they were on straight femoral traction in order to monitor the rotation. Details of the subject groups are given in Table 2.1. Whilst measuring the rotation of the femur the alignment of the fracture and the progress of callus formation were also assessed.

Table 2.1. *Age and sex distribution of the three groups of cases studied*

No. of cases	Sex	Age range (in years)	Average age (in years)
30 volunteers	9 male 21 female	18–50	28.1
25 patients (healed femoral fracture)	19 male 6 female	17–45	30.0
10 patients (fresh femoral fractures)	8 male 2 female	17–42	25.3

Experimental technique

A standard ultrasound machine (Technicare EDP 1000 with calibrated velocity 1540 m/s with probe frequency 5-7.5 MHz) was used. The volunteers and patients were scanned on an ordinary table or bed. Probe to skin contact was made using Aquasonic Gel. Both limbs were placed in neutral rotation, and the probe was used to scan two sections: one distally through the condylar region of the femur, about 2.5 cm proximal to the joint line of knee and the other at the level of the greater trochanter passing the probe transversely towards the hip joint. The correct level is difficult to define clinically but the shape of the head, neck, and trochanteric region of the femur can be easily recognized with the ultrasound probe. The procedure can be repeated on the other limb, provided that the limb is not moved between the knee and the hip scan. Transcondylar and femoral neck axes were drawn on these two scan pictures and the

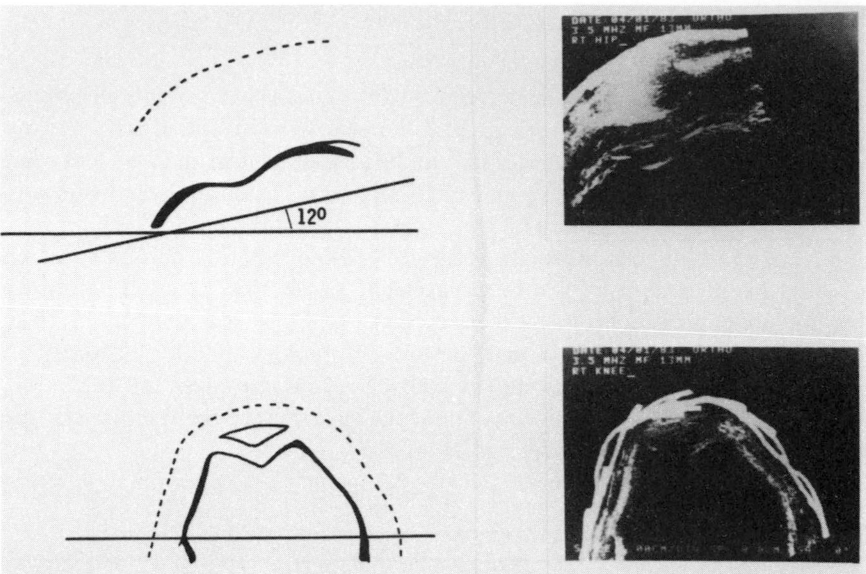

Fig. 2.1. An example of the ultrasound display and the method of measuring the angle of anteversion.

angle of femoral anteversion or torsion was then measured between the two axes (Fig. 2.1).

When scanning the 10 patients on traction, no attempt was made to alter the attitude of the hip prior to scanning. Scanning was repeated at weekly intervals starting from the second week during the traction period to monitor the rotational element at the fracture site to keep the difference between the sides within 6°. The rotation at the fracture site was controlled by applying vertical traction to one end of the tibial pin. The fracture site was also scanned both axially and transversely.

RESULTS

The results of 30 volunteers with no hip abnormality and seemingly normal walking pattern are shown in Table 2.2. The average difference between right and left hip was +2.7° and none had more than 6° difference.

A different state of affairs was noted in 25 patients with healed unilateral femoral shaft fractures (Table 2.3). Ten patients had more than 10° difference between the hips with a range of 10-20°. The average difference in these 10 cases was +13.7°. Twelve cases had 6° or less difference including only two cases in which there was the same reading on both sides. The average difference in these 25 cases was +8.25°.

Table 2.2. *The normal variation in the 30 volunteers with normal walking pattern, between the right and left hips of the anteversion angle*

Difference of angle	No. of volunteers
0°	10
2°	6
4°	7
6°	7
Total	30

Average difference between the sides +2.7°.

Table 2.3. *To show the difference in the 25 cases of healed femoral fractures in anteversion angle on accepted conservative fracture management*

Difference of angle between the sides	No. of patients
0°	2
2–6°	10
8–10°	4
11°	1
12°	4
14°	2
20°	2
Total	25

Average difference between the sides +8.25°.

The aim of prospective study was to see if it were possible to equalize the anteversion angles by modifying the traction pull and we found that up to three weeks post-injury the rotation can be altered by rotatory traction.

DISCUSSION

Small angles of rotatory deformity of the femora can be apparently accommodated by the patient and give no cosmetic or functional disability. However, Reikeras and Hoiseth (1982) concluded that after a radiological study of 44 patients, an increased anteversion angle may lead to osteoarthritis of the hip. This was confirmed in a study using computed axial tomography (Reikeras *et al.* 1983). We put forward this method of monitoring rotatory deformity in the hope of avoiding these complications. Unfortunately, the incidence of

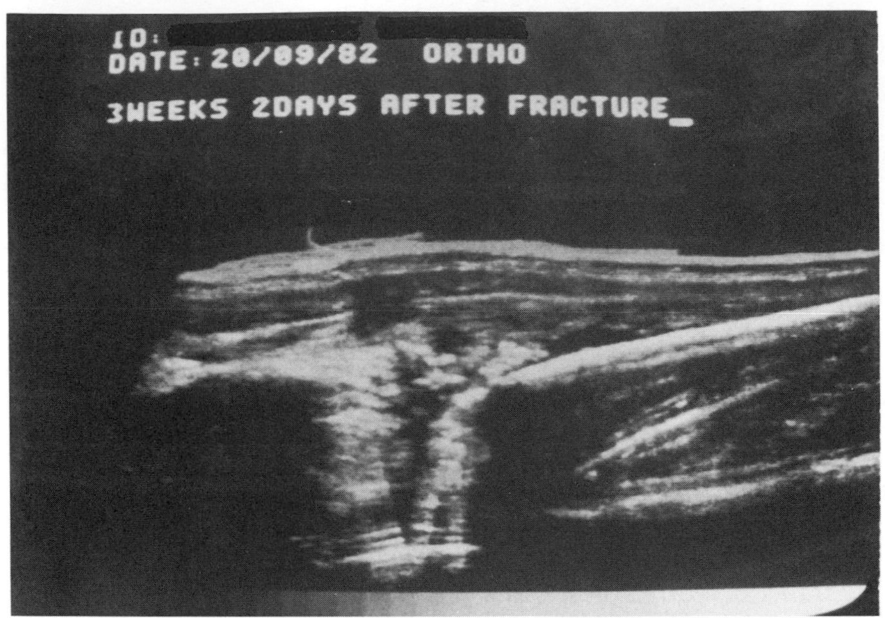

Fig. 2.2. Ultrasound display of callus formation at three weeks.

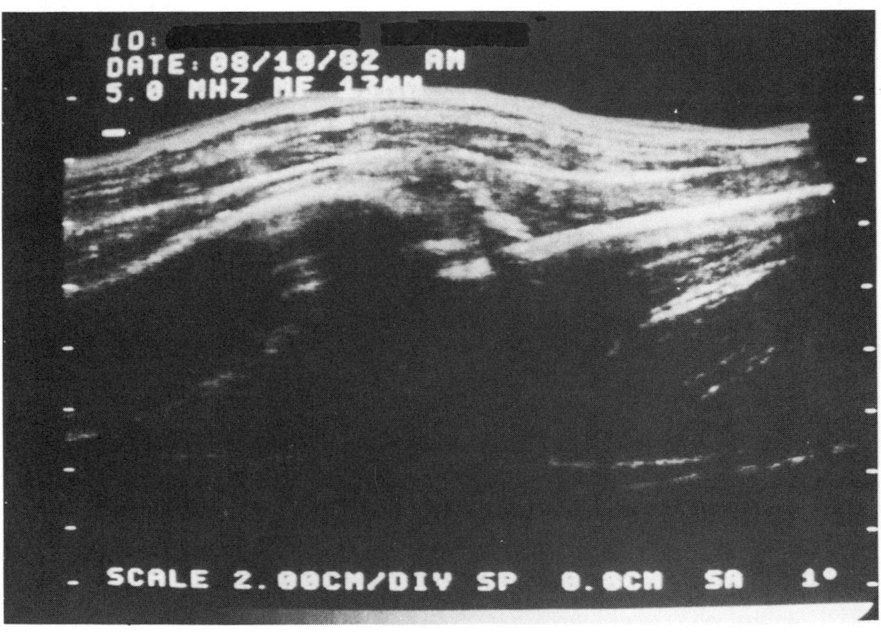

Fig. 2.3. Ultrasound display of callus formation at six weeks.

osteoarthritis of the hip following fracture of the shaft of femur is not known, no study having been revealed in a literature search.

In the prospective trial, 10 patients were monitored with ultrasound for equalization of the femoral anteversion angle, and the angles have been kept within 6° of the normal side (+ or −3°). Adjustment can be made up to three weeks. Rotatory traction does not add to the patient's discomfort. In addition to the measurement of femoral anteversion angle whilst the patient is on traction, an opportunity presents to see what changes occur at the fracture site; further work is in progress quantitatively to assess the soft callus in correlation with the radiological appearance (Figures 2.2 and 2.3); it may be possible to assess fracture union and alignment using ultrasound and to assess soft tissue repair. If it is confirmed that an increased anteversion angle predisposes to osteoarthritis, then it behoves the clinician to maintain the patient's anteversion angle in cases of fracture.

Acknowledgements

The authors are grateful to Mr T. O'Neil of the Ultrasound Department, Mansfield and District General Hospital for his co-operation and help in this study. Thanks are also due to Miss E. A. Brown, Mrs S. Ounsworth, and Mrs J. Parker for their help throughout the study.

REFERENCES

Moulton, A. and Upadhyay, S. S. (1982). A direct method of measuring femoral anteversion using ultrasound. *J. Bone Jt Surg.* **64B**, 469–72.

Reikeras, O., Bjerkreim, I., and Kolbenstvedt, A. (1983). Anteversion of the acetabulum and femoral neck in normals and in patients with osteoarthritis of the hip. *Acta orthopaed. Scand.* **54**, 18–23.

— and Hoiseth, A. (1982). Femoral neck angles in osteoarthritis of the hip. *Acta orthopaed. Scand.* **53**, 781–4.

3 MADAMS-Percussion Technique: A Non-Invasive Vibration Analysis Technique to Monitor Fracture Healing

M. CORNELISSEN, G. VAN DER PERRE, F. BURNY, AND M. MULIER

INTRODUCTION

Results of '*in vivo* vibration' analysis of one point on the ulna have been published previously, e.g. resonant frequency measurements (Jurist 1970) and mechanical impedance measurements (Thompson 1973). However, it was pointed out by Doherty *et al.* (1974) that results of one-point vibration response measurements can only be interpreted in terms of mechanical properties when the excitated vibration modes are correctly identified.

In the MADAMS-percussion technique this identification of modes is a primary goal because it opens the possibility of correlating vibration analysis results with mechanical properties. When a structure is excited, by any kind of force, it starts vibrating. This vibration is a superposition of some modes (vibration shapes) which are characteristic for the structure (Figs. 3.1 and 3.2). Each mode has its own vibration frequency (natural frequency) and damping coefficient. Studies carried out during the last five years on dry and freshly excised human tibiae and *in vivo* measurements revealed a lot about the vibrations of the human tibia (Van der Perre *et al.* 1981).

To correlate vibration with mechanical characteristics the knowledge of the vibration mode associated with a measured natural frequency is necessary. A **torsion** mode will be correlated with torsion stiffness, and a **single-bending** mode with bending stiffness (transverse stiffness). A **double-bending** mode will also be correlated with bending stiffness, but this correlation will differ from the previous one.

The vibration analysis methods can be divided into the one-point and the multiple-point methods. Only the modal analysis (multiple-point method) gives as results the natural frequency, mode shape, and damping coefficient. The one-point techniques make an assumption about the mode shape. When the structure is well known this can be done if, in a previous study, the modes were identified and localized in regions of the frequency domain.

When the structure is unknown or changing, e.g. in a healing tibia, it is very difficult to make such an assumption because the modes and frequency regions are changing too.

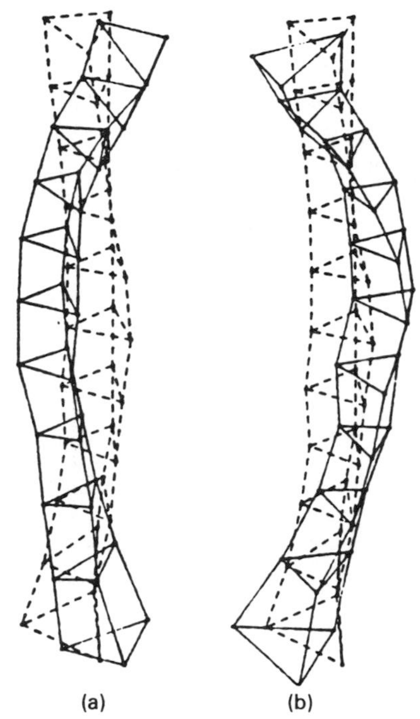

Fig. 3.1. Single-bending modes of dry tibia. (a) Direction of minimum EI (x) 497 Hz. (b) Direction of maximum EI (y) 649 Hz.

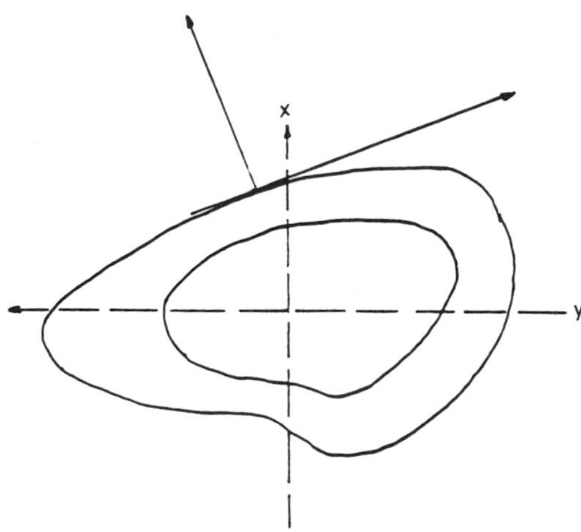

Fig. 3.2. Directions of principal axes (x,y) in a tibial cross-section.

METHOD

The acronym 'MADAMS' was designed to characterize an approach which has been developed in our laboratory. It contains two basic elements:

1. Modal Analysis.
2. Dynamic Anatomical Modelling for Stability.

Modal analysis (Fig. 3.3)

Modal analysis is an advanced experimental procedure for the analysis of the vibration behaviour of a structure in mechanical engineering.

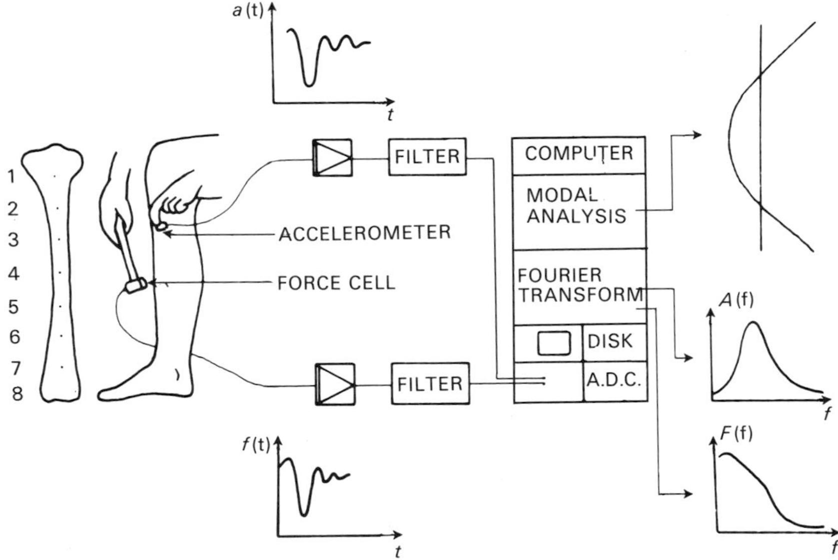

Fig. 3.3. Modal analysis set up.

What is measured?

The input force at the point of excitation (using a load cell in the impact hammer) and the motion (using an accelerometer) of the structure in a set of points.

What is calculated?

First, the transfer functions to eliminate the influence of the input force. Then, these transfer functions are analysed and from the combination of transfer functions the mode shapes, natural frequencies, and damping coefficients are reconstructed by modal analysis computer algorithms.

We started to use this MA technique on human bones. The vibrations are excited by hammer impact. For *in vivo* measurements this hammer impact is exerted upon the medial malleolus.

Dynamic anatomic modelling for stability

We developed mathematical models for a healing bone (Cornelissen *et al.* 1982) to predict: (i) the vibrational behaviour; (ii) the axial stability (Fig. 3.4); (iii) the transverse stiffness (Fig. 3.5). The measurement (and subsequent modal analysis) is made on the patient's healing bone and his opposite (reference) bone. The frequencies of the single bending modes are selected and the 'dynamic ratio' is evaluated.

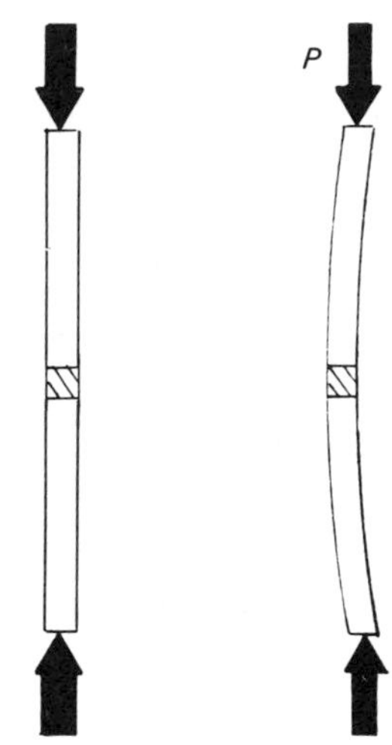

Fig. 3.4. Buckling strength (p).

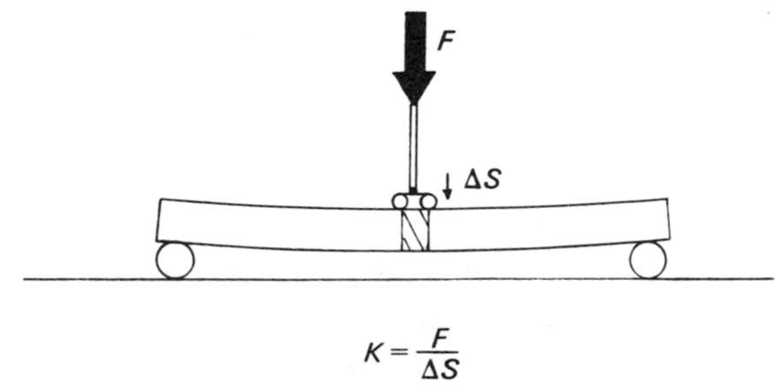

Fig. 3.5. Transverse stiffness (k).

The ‘dynamic ratio’ (DR) is defined as:

$$DR = \frac{f_H}{f_R}$$

where f_H is frequency of the single bending mode of the healing bone, and f_R is the frequency of the single bending mode of the reference bone.

The ‘ratio of the buckling strength’ (RBS) is defined as:

$$RBS = \frac{p_{cr.H}}{p_{cr.R}}$$

where $p_{cr.H}$ is the buckling strength of the healing bone and $p_{cr.R}$ is the buckling strength of the reference bone.

The ‘transverse stiffness ratio’ (TSR) is defined as:

$$TSR = \frac{k_H}{k_R}$$

where k_H is the transverse stiffness (Fig. 3.5) of the healing bone and k_R is the stiffness of the reference bone (same loading configuration).

Using these models, quantitative relations are set up between DR on the one hand and the mechanical properties RBS and TSR on the other hand.

$$\begin{aligned} RBS &= f_1\,(DR, \text{parameters}) \\ TSR &= f_2\,(DR, \text{parameters}) \end{aligned}$$

Procedure

Measurements were made at intervals following fracture. At the moment, the MA is not yet automatic and consequently it is very time consuming. The results, however, are quantitative indices with the physical meaning of a stability and stiffness ratio.

No special qualifications are required to make the measurements. However, at the moment, only people with knowledge of the modal analysis software can run the analysis. We intend to speed up the procedure as soon as we have enough experience and knowledge to do this without loss of accuracy.

RESULTS

Clinical studies

We present the results of a series of five patients. Tables 3.1-3.5 give the numerical values of DR, TSR, and RBS for the five patients. All the patients had a single fracture of the tibia; the clinical history is given in Table 3.6. Figures 3.6-3.9 give the time course of the studies on patients 1-4. The fifth patient, a non-healing case was measured only twice, and we did not draw a curve for this case.

Table 3.1. *Patient 1*

Time	Measurement results			
Days after operation	Static	MADAMS		
	100 % – Strain %	DR %	TSR %	RBS %
55	66.5	46	18	22
76	77	56	27	33
104	79	75	50	59
111	77	75	50	59
125 (EOF)	81	79	56	65
132		89	75	82

EOF = End of fixation.

Table 3.2. *Patient 2*

Time	Measurement results			
Days after operation	Static	MADAMS		
	100 % – Strain %	DR %	TSR %	RBS %
43	27	–	–	–
57	30	53	17	24
78	49	60	23	31
106 (EOF)	52	62	25	34
134		1107	131	116

EOF = End of fixation.

Table 3.3. *Patient 3*

Time	Measurement results			
Days after operation	Static	MADAMS		
	100 % – Strain %	DR %	TSR %	RBS %
89	19	62	21	30
96	23	61	20	29
110	22	66	24	35
159 (EOF)	–	82	42	60
187	–	80	39	57

EOF = End of fixation.

Table 3.4. *Patient 4*

Time	Measurement results			
Days after operation	Static	MADAMS		
	100 % − Strain %	DR %	TSR %	RBS %
87	43	60	32	38
115		62	34	41
129	67.5	60	32	38
143	64	57	29	35

Table 3.5. *Patient 5*

Time	DR	TSR	RBS
X	85%	34%	56%
$X + 2\frac{1}{2}$ month	91%	51%	71%

Table 3.6.

Patient 1
Age: 16 year
Sex: male
Accident: motorbike
Fracture: Tibia, 17 cm above malleolus; fibula: proximal
Treatment: external fixation
End of fixation: 125 days (four months)
Remarks: faster healing after removal of fixation

Patient 2
Age: 16 year
Sex: male
Accident: motorbike
Fracture: Tibia, 12.5 cm above malleolus
Treatment: external fixation
End of fixation: 106 days ($3\frac{1}{2}$ months)
Remarks: very fast healing after fixation has been taken away

Patient 3

Age: 40 year
Sex: male
Accident: fall
Fracture: Tibia, 11 cm above malleolus
Treatment: external fixation
End of fixation: 159 days ($5\frac{1}{3}$ months)
Remarks: no evolution after fixation has been taken away

Patient 4

Age: 19 year
Sex: male
Accident: car
Fracture: open, tibia, 18 cm above malleolus; fibula: yes
Treatment: external fixation
End of fixation: non-healing
Remarks: pseudo arthrosis

Patient 5

Age: 40 year
Sex: male
Accident: car
Fracture: Tibia, 8 cm above malleolus; fibula: proximal
Treatment:

6/78	osteosynthesis with plate and screws
2/79	osteosynthesis material removed external fixation
11/79	external fixation removed patella tendon-bearing apparatus
12/79	walking cast
14/1/80	bandage
3/80	electromagnetic bone stimulation (EBI, external, magnetic field) cast
9/80	end of EMBS PTB apparatus
7/81	no PTB apparatus any more working again
12/81	pain again radiolucent line appeared EMBS PTB apparatus
13/5/82	MADAMS-percussion technique cast
8/82	MADAMS-percussion technique
10/82	PTB apparatus
2/83	radiographic: positive evolution

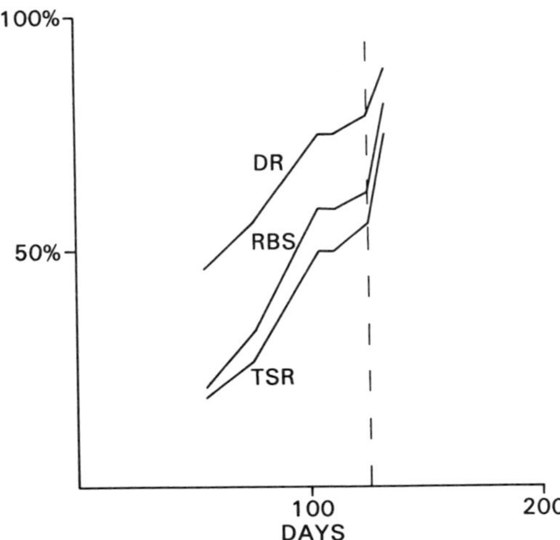

Fig. 3.6. Patient 1: curves DR, RBS, and TSR.

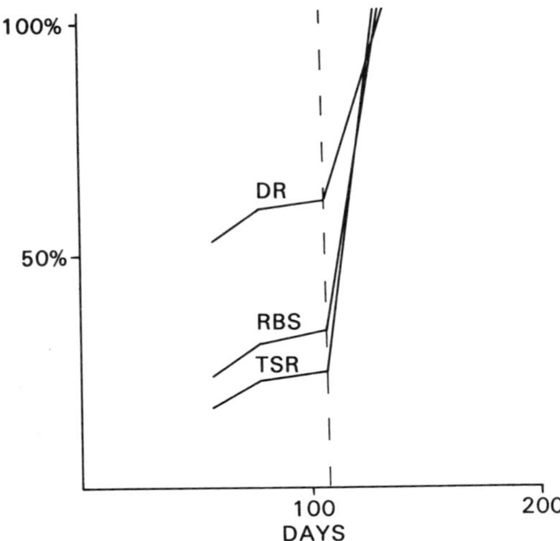

Fig. 3.7. Patient 2: curves DR, RBS, and TSR.

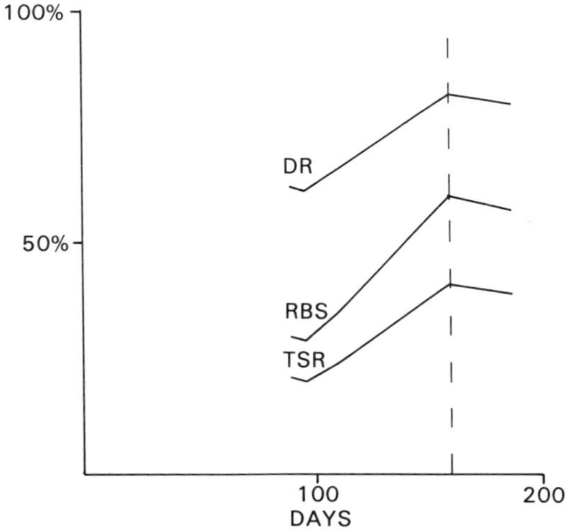

Fig. 3.8. Patient 3: curves DR, RBS, and TSR.

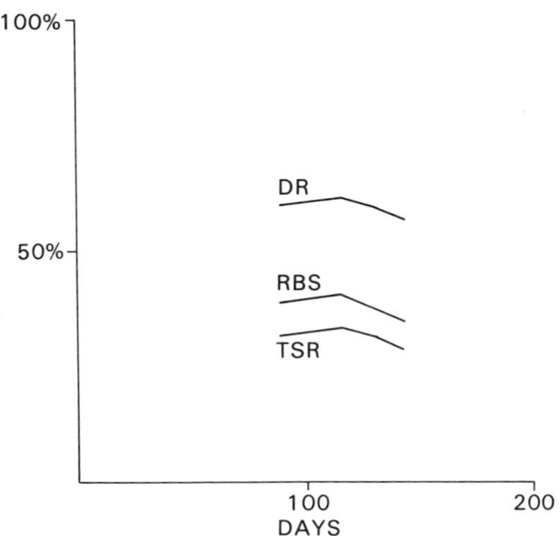

Fig. 3.9. Patient 4: curves DR, RBS, and TSR.

COMPARISON: MADAMS-PERCUSSION TECHNIQUE VERSUS STATIC STRAIN GAUGE TECHNIQUE

Owing to technical and practical circumstances the series of patients measured with both techniques is limited. Only three cases, included one non-healing, are presented.

The deformation of the fixation bar, loaded by the leg's own weight, is measured using a strain gauge technique (Burny 1979). The maximum deformation is taken as a reference (100 per cent). The other measurements are related to this maximum. So, a curve of decreasing deformation is obtained.

$$\text{strain}\,\% = \frac{\text{strain}_{(t)}}{\text{strain}_{\max}}$$

Because we compare this curve with the increasing DR-curve of the MADAMS-percussion technique, we define the strain-complement (STC):

$$\text{STC} = 100\,\% - \text{strain}\,\%.$$

This STC-curve is also an increasing one. Figures 3.10–3.12 give both curves for patients 1, 2, and 4. The correlation coefficients between both results were calculated.

$$R_{xy} = \frac{\Sigma(x-\bar{x})\,(y-\bar{y})}{[\Sigma(x-\bar{x})^2\,\Sigma(y-\bar{y})^2]^{\frac{1}{2}}}$$

$$-1 \leqslant R_{xy} \leqslant 1.$$

When there is a linear relationship $|R_{xy}| = 1$.
When there is no linear relationship $0 \leqslant |R_{xy}| < 1$.

Patient 1 : R_{xy} = 0.89 (7 points).
Patient 2 : R_{xy} = 0.998 (3 points).
Patient 4 : R_{xy} = −0.10 (5 points).

DISCUSSION

The number of patients monitored so far is too limited to draw statistical conclusions. However, the fact that in two cases we observed a faster healing after the removal of the fixation, can be important if this can be confirmed statistically.

Another important fact is that the curve for non-healing is clearly different from normal healing. It is also possible to determine whether electromagnetic bone stimulation has a positive effect or not.

The correlations for the two normal-healing patients were very good although the two measurements have a different nature. We planned our research in the direction of standardization of the measurement configuration so that simplification of the analysis will be possible. When that is done a large number of patients can be measured and statistically relevant data will be available.

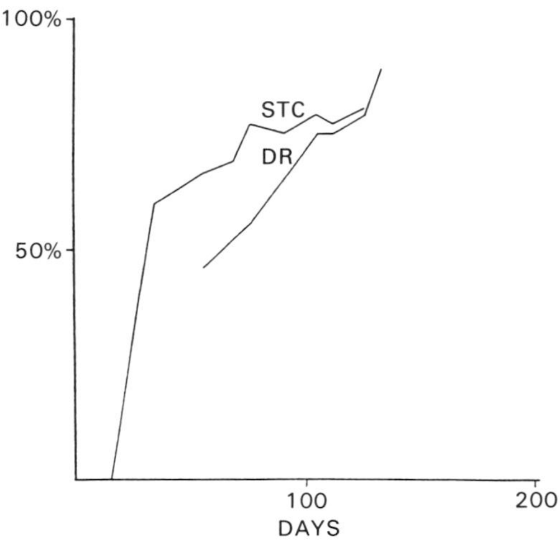

Fig. 3.10. Patient 1: comparison of DR and STC.

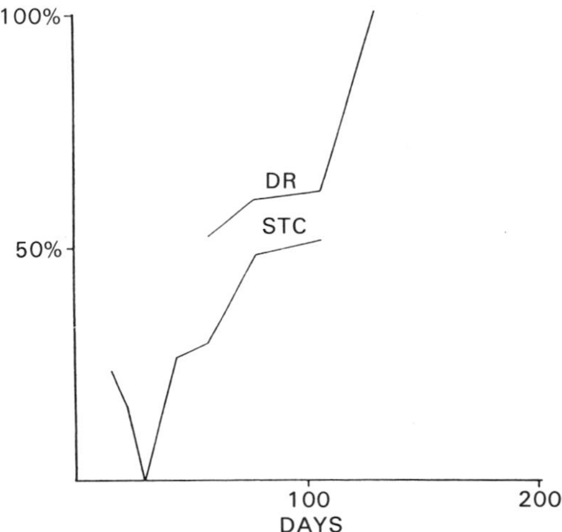

Fig. 3.11. Patient 2: comparison of DR and STC.

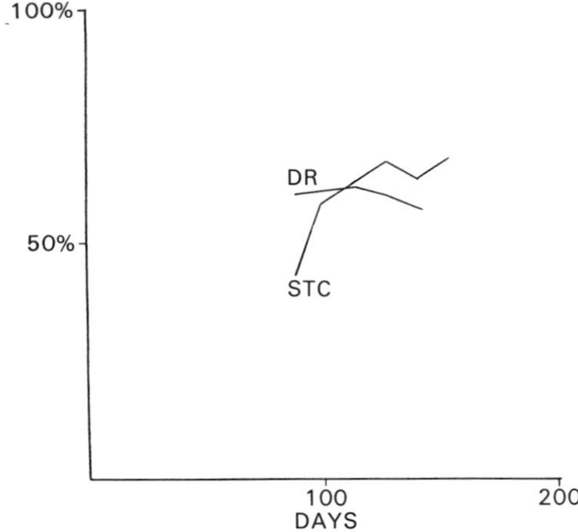

Fig. 3.12. Patient 4: comparison of DR and STC.

REFERENCES

Burny, F. (1979). Strain gauges measurements of fracture healing. A study of 350 cases. In *External fixation: the current state of the art* (ed. E. F. Brooker and C. C. Edwards) pp. 371–82. Williams & Wilkins, Baltimore.

Cornelissen, M., Cornelissen, Ph., and Van der Perre, G. (1982). A dynamic model for a healing fractured tibia. In *Biomechanics: principles and applications* (ed. R. Huiskes, D. H. Van Campen, and J. P. de Wijn) pp. 213–18. Martinus Nijhoff, The Hague.

Doherty, W. P., Bovill, E. G., and Wilson, E. L. (1974). Evaluation of the use of resonant frequencies to characterise physical properties of human long bones. *J. Biomechan.* **7**, 559.

Jurist, J. M. (1970). *In vivo* determination of the elastic response of bone. I. Method of ulnar resonant frequency determination. *Phys. Med. Biol.* **15**, 417.

Thompson, G. T. (1973). *In vivo* determination of bone properties from mechanical impedance measurements. Aerospace Med. Ass. Annual Science Meeting Abstracts, 7–10 May, Las Vegas, Nevada.

Van der Perre, G., Van Audekercke, R., Vandecasteele, J., Martens, M., and Mulier, J. C. (1981). Modal analysis of human tibiae. In Proceedings ASME-Biomechanics Symposium, Boulder, June 1981, AMD-Vol. 43, pp. 173–6.

4 The Monitoring of Fracture Stiffness with External Fixation

M. EVANS, J. GWILLIM, DEREK HARRIS, AND K. E. TANNER

INTRODUCTION

The clinical assessment of fracture stiffness during rehabilitation depends on the surgeon's manual tests for instability, the patient's response in terms of pain and confidence in load bearing, and on radiographic evidence of callus and primary bone union.

None of this evidence is readily quantifiable, so no objective comparison of differing regimens of fracture management is possible, unless other measures are available. Additionally, without objective measurement, the degree of healing cannot be accurately determined to assist the surgeon in deciding when the fixator may be safely removed.

A system for the measurement of fracture stiffness using a strain-gauge transducer on a rigid type of single-sided fixator has been developed, and patient tests have been carried out over the last two years to measure progressive stiffness during healing. Standard tests to stress the fracture and measure the external load applied are used, and different fracture management routines are compared. Although variation in the severity of the initial trauma and the type of fracture cause considerable differences in the rate of healing, three patterns of healing with time are described.

METHOD

When a fractured limb is supported through bone screws to a unilateral fixation frame and the limb is subjected to external loading, the distribution of the load will be shared between the limb and the frame dependent upon their relative stiffnesses, and the stiffness of the screw fixation in the bone and in the clamping system between screw and frame. As the fracture stiffness increases with healing so the proportion of load carried by the fracture will increase, and the proportion carried by the frame, decrease.

Because of the offset of the fixator from the axis of the bone, the loads in the fixator frame will be predominantly bending and torsional.

Conventionally, to measure bending in a beam, strain gauges are bonded onto the metal structure. This system is not practical for a fracture fixation frame in any large-scale patient-testing study. The strain gauge would be subject to mechanical and environmental damage during the surgical operation and probably also when the patient is walking and bathing, etc. More importantly, individual calibration of the strain gauge before loading would be difficult and subsequent calibration checks for drift during the ensuing weeks impossible.

To overcome these problems a clamp-on load-monitoring transducer has been developed. This is fitted to the fixator of patients attending regular fracture clinics. The transducer is shown schematically in Fig. 4.1. Basically it consists of

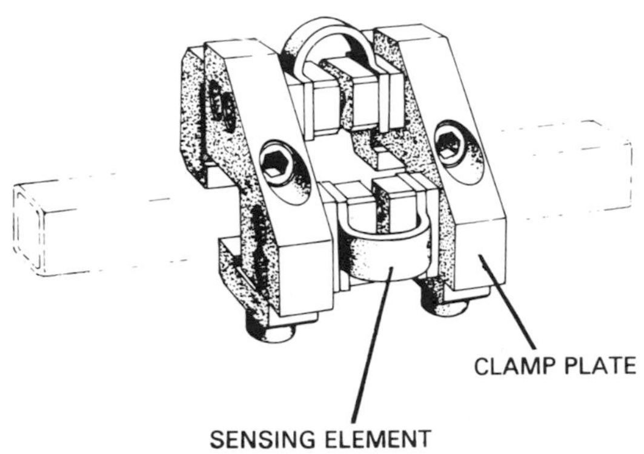

Fig. 4.1. Clamp-on load monitoring transducer.

two clamp-plates bridged by two strain-gauge sensing elements arranged at 90° to each other. This device measures bending in two planes and torsion, with a linear response to within 1 per cent, and a cross-talk between channels of less than 5 per cent in bending and 15 per cent in torsion. The removal and replacement of the transducer on the fixator produces repeatable results to within 5 per cent, and takes only three minutes.

At the start of the clinic the strain gauge unit is mounted on a standard frame and calibrated. It is then attached to the frame of the fixator on the patient in a standardized position.

The output signals from the transducer when monitoring fracture healing are normally less than 100 microstrain and to record these low signals a three-channel amplifier has been built. Care was taken to eliminate signal noise to provide good resolution, and a stable gain was achieved.

PATIENT TESTING

With tibial fractures three tests are carried out on patients, commencing as soon as possible following the application of the fixator, at weekly or two-weekly intervals. Two of these tests involve axial loading of the tibia. In one case the patient is seated with the foot resting on a force plate, with a gradually increasing external load applied to the knee; in the second situation the patient is standing with the foot of the injured limb resting on the force plate. The body weight is gradually transferred to that limb up to the limits of comfort. Although both tests give an indication of increasing axial stiffness as healing occurs, the results are influenced by the degree of alignment of the fracture and by muscle action. The third test measures the bending and shear stiffness of the fracture without any axial load on it. In this test, with the patient on a couch (Fig. 4.2), the difference is measured between the transducer reading with the leg unsupported, and with the leg supported at the heel. On all occasions the rotation of the leg is standardized.

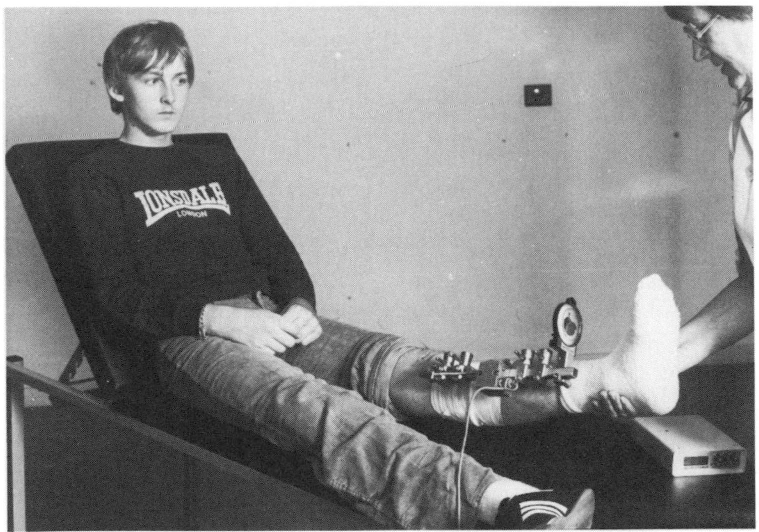

Fig. 4.2. Subject undergoing straight leg raise test with heel supported.

Data from the three strain-gauge transducer readings, and where appropriate from the floor-mounted force plate, are processed by a PDP-11/34 computer and indices of fracture stiffness are calculated. Stiffness indices measured by the three tests differ owing to the effect of the geometry of the fracture and the differing forces produced.

A progress chart is produced for each patient. This shows the change of stiffness index as healing progresses. Figure 4.3 is a typical output for a patient. Also shown on this chart is the percentage body weight that the patient was able to support on this limb as healing progressed; this is primarily a measure of confidence in the fixator by the patient.

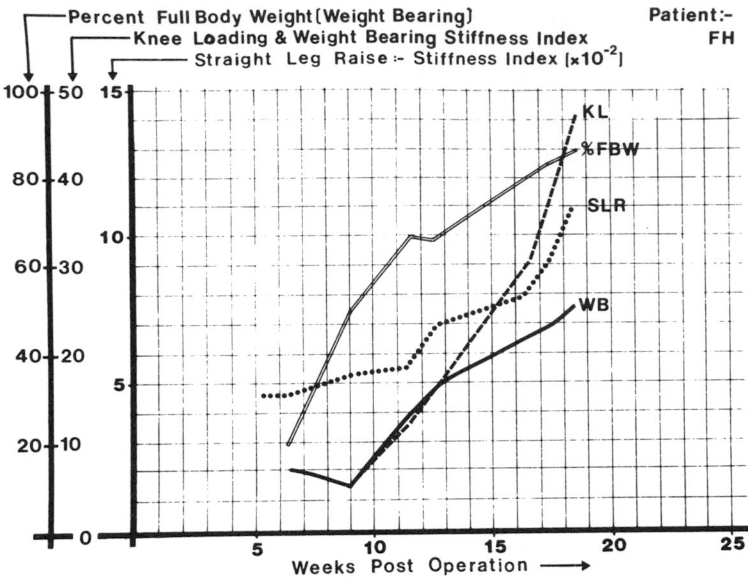

Fig. 4.3. Progression of four measured parameters in one subject: percentage full-body weight (%FBW); knee-loading stiffness index (KL); weight-bearing stiffness index (WB); and straight leg raise stiffness index (SLR).

RESULTS FROM MEASURING FRACTURE MONITORING

Experience gained over the past 18 months has shown that the bending/shear test, as performed with the straight leg raise, gives the most reliable correlation with the clinical assessment of fracture healing, and this test is the basis of the group study below.

The results from fracture monitoring can be divided into three groups of patients, each group having a distinct pattern and rate of healing.

Figure 4.4 shows patients with a mid-tibial shaft fracture treated by external fixation. Healing is uneventful with a steady increase in stiffness being recorded. The stiffness index for this group increased by 100 per cent over a 24-week period, with all fractures going on to good union.

The second group (Fig. 4.5), with mid-shaft fractures, had received a late bone graft owing to clinical indications of delayed union. No clear healing trend can be seen, nor is the contribution made by the bone graft distinct.

Group 3 patients, with severe mid-shaft tibial fractures and some with appreciable loss of bone with soft-tissue damage, received bone grafting within three weeks of the application of the fixator. With this group the rate of healing was very rapid, the stiffness index increasing by 100 per cent within 12–17 weeks.

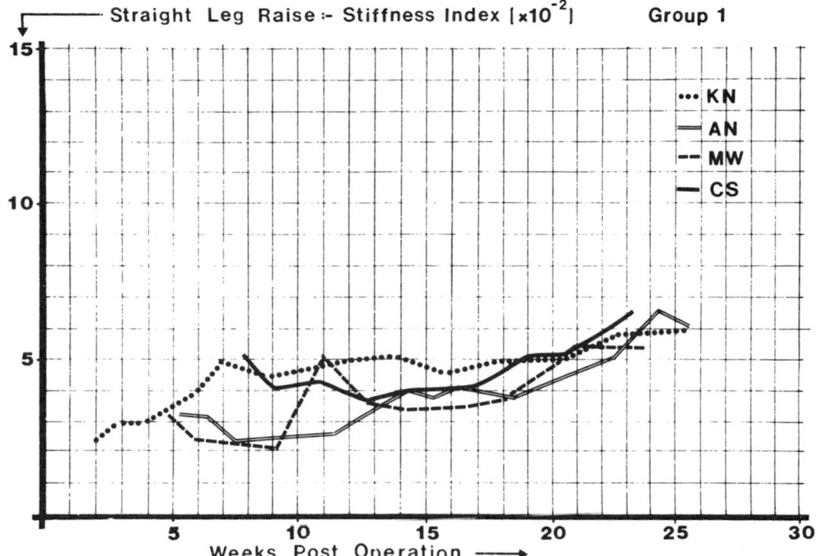

Fig. 4.4. Progression of straight leg raise stiffness index in four subjects (Group 1).

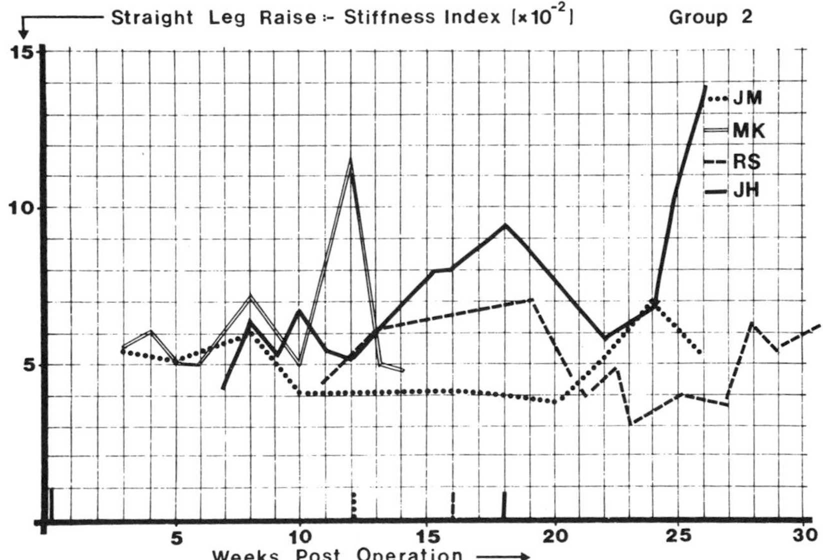

Fig. 4.5. Progression of straight leg raise stiffness index in four subjects (Group 2). Vertical lines above bottom axis are dates of bone grafting.

With this group of patients there are a number of observations that can be made from the progress chart (Fig. 4.6). The initial rate of healing, up to 10 weeks, is very similar to that of Group 1. Then, at between 10–12 weeks, there is a rapid increase in the rate of healing. The overall 100 per cent increase in stiffness index is achieved in less than 17 weeks, as compared to 25 weeks from Group 1. Assessment of these patients by normal clinical methods confirmed this improvement which allowed the fixators to be removed at between 12 and 22 weeks.

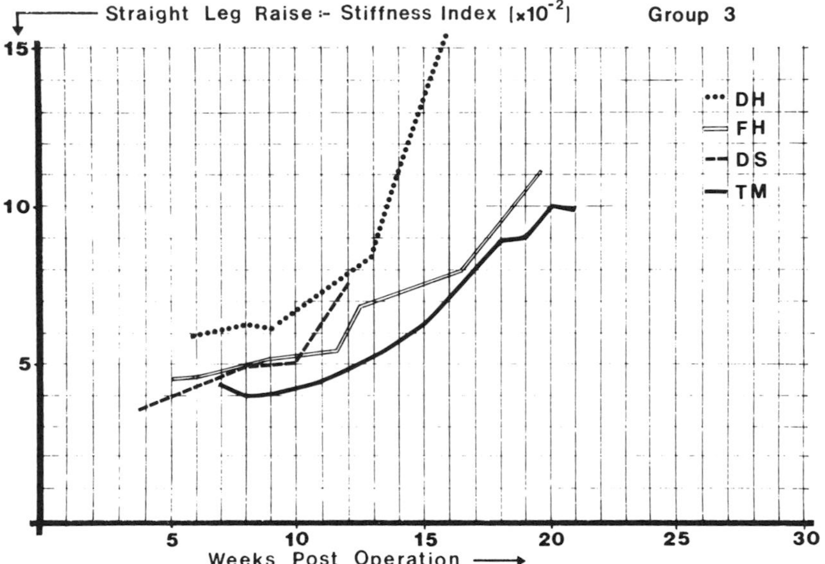

Fig. 4.6. Progression of straight leg raise stiffness index in four subjects (Group 3).

DISCUSSION AND CONCLUSIONS

The need for some form of non-invasive objective assessment of fracture healing is recognized, and the development of the fracture-monitoring transducer for use with fractures supported by external fixation has proved to be a useful tool for this purpose.

The ability to clamp the transducer onto a fixator column means that a periodic check can be made of the system calibration. It can be used on any patient at any time, making it possible to monitor fracture healing on a weekly or even daily basis. The system provides objective information for the assessment of a particular fracture, and indications for subsequent management.

Comparisons can be made between individual patients and patient groups to provide a better understanding of fracture healing and the effect of a particular form of fracture management. For example, in the study presented here there is a good indication that early bone grafting is effective in promoting rapid bone healing. Also, that in Group 3 the rate of healing increases rapidly at about 10 weeks. A better understanding of the causation of this improvement would undoubtedly assist in the general management of fracture healing.

5 Nail Plates Instrumented with Strain Gauges: Review of the First 100 Cases of Fractures of the Upper Extremity of the Femur

F. BURNY, M. DONKERWOLCKE, R. BOURGOIS, AND O. SARIC

INTRODUCTION

The breakage of implants used for fixation of fractures of the proximal femur is a frequent occurrence and severely impairs recovery of patients (Ramadier *et al.* 1956; Decoulx and Lavarde 1969; Panda 1974). Since 1965, we have used an extensometry technique (strain gauges) to assess fracture healing; this technique was extended to the measurement of deformation of nail plates in 1975. *In vivo* measurement of the deformation of implants using strain gauges produces useful information on the actual mechanical characteristics of the bone-callus-implant complex. The technique is used to monitor the deformation of nail plates during rehabilitation of patients who have previously suffered a fracture of the upper end of the femur.

MATERIALS AND METHOD

We generally use regular MacLaughlin nail plates instrumented with one strain gauge (TML PS3 120 ohm) longitudinally bonded on the lateral aspect of the plate. In five cases we used an experimental monobloc nail plate with a special accommodation to protect the transducer (Fig. 5.1). The type of implant and the indications are presented in Table 5.1.

Previous studies (Burny *et al.* 1972; Moulart 1974-75) and analysis of implant failures have shown that the maximum stress appears at the nail-plate junction for trochanteric fractures and at the level of fracture line for sub-trochanteric fractures. At the time of the operation an appropriate nail plate is selected with correct alignment of the strain gauge.

During fracture healing the implant will support part of the mechanical loading, depending on the characteristics of the fractures, on the rate of healing, and on the stability of the mounting.

The readings are obtained using a transcutaneous lead.

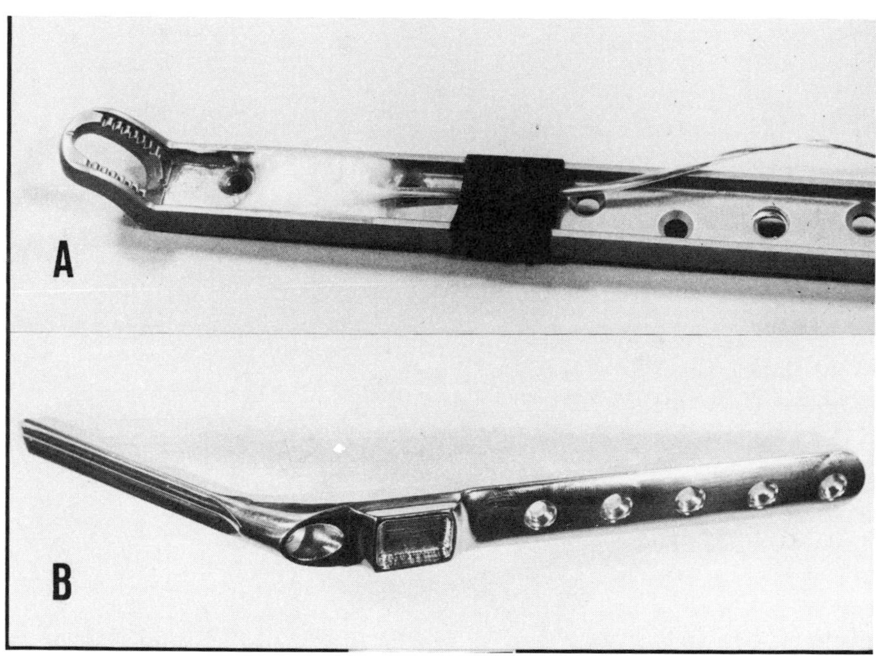

Fig. 5.1. MacLaughlin nail plate with strain gauge (A). Experimental monobloc nail plate (B) with a facet for the strain gauge.

Table 5.1. *Type of implant*

Fracture	ML 7 holes	ML 9 holes	ML 12 holes	Monobloc
Cervical	12	1		1
Pertroch. simple	28	2		4
Pertroch. comminuted	12	3		
Subtrochanteric	5	8	24	
Total	57	14	24	5

Because of the shape of the MacLaughlin plate the recorded deformations are not representative of the maximum stress of the implant. On a section of a plate (Fig. 5.2) we calculated that the readings correspond to about 0.4 of the maximum stress at the extremities of the flanges. We thus have to multiply the reading by a factor of approximately 3. The corrected value of the deformation has to remain under a threshold depending on the characteristics of the material.

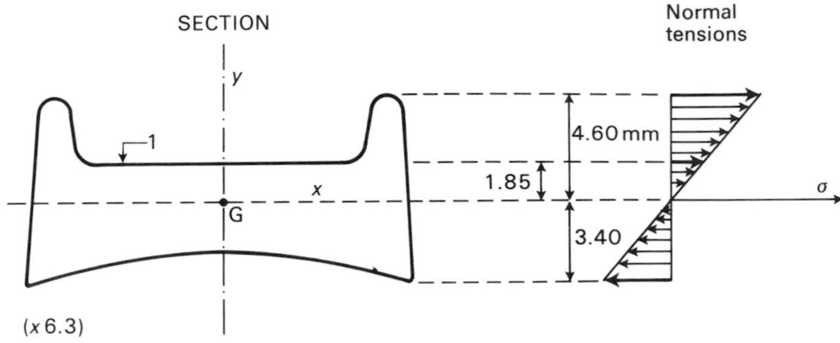

Fig. 5.2. Section of a MacLaughlin plate; calculation of maximum stress.

If we consider the equation $\sigma = \Delta 1 . E$ we have

$$50.4 = \Delta 1.10^{-6} \times 20.4\ 10^3 \times 3$$

where 50.4 kg/mm^2 is the elastic limit of the vitallium; 20.4×10^3 kg/mm^2 is the Young's modulus of the vitallium; and 3 is the correction factor. Thus

$$\Delta 1 = 1214.10^{-6}$$

Deformations greater than 1200 μs are in the plastic range of deformation of the vitallium and can be responsible for fatique failure.

RESULTS

The general characteristics of the patients are presented in Table 5.2. Measurements of implant deformation were recorded during rehabilitation exercises (Fig. 5.3). We were also interested in the rate of soft-tissue complications (Table 5.3).

Table 5.2. *General characteristics of the patients*

	No.	Mean age
Men	42	57.1
Women	58	74.0
Total	100	

Table 5.3. *Complications related to soft tissue*

No complications	87
Haematoma	3
Infected haematoma	1
Deep infection	5
Unknown	4
Total	100

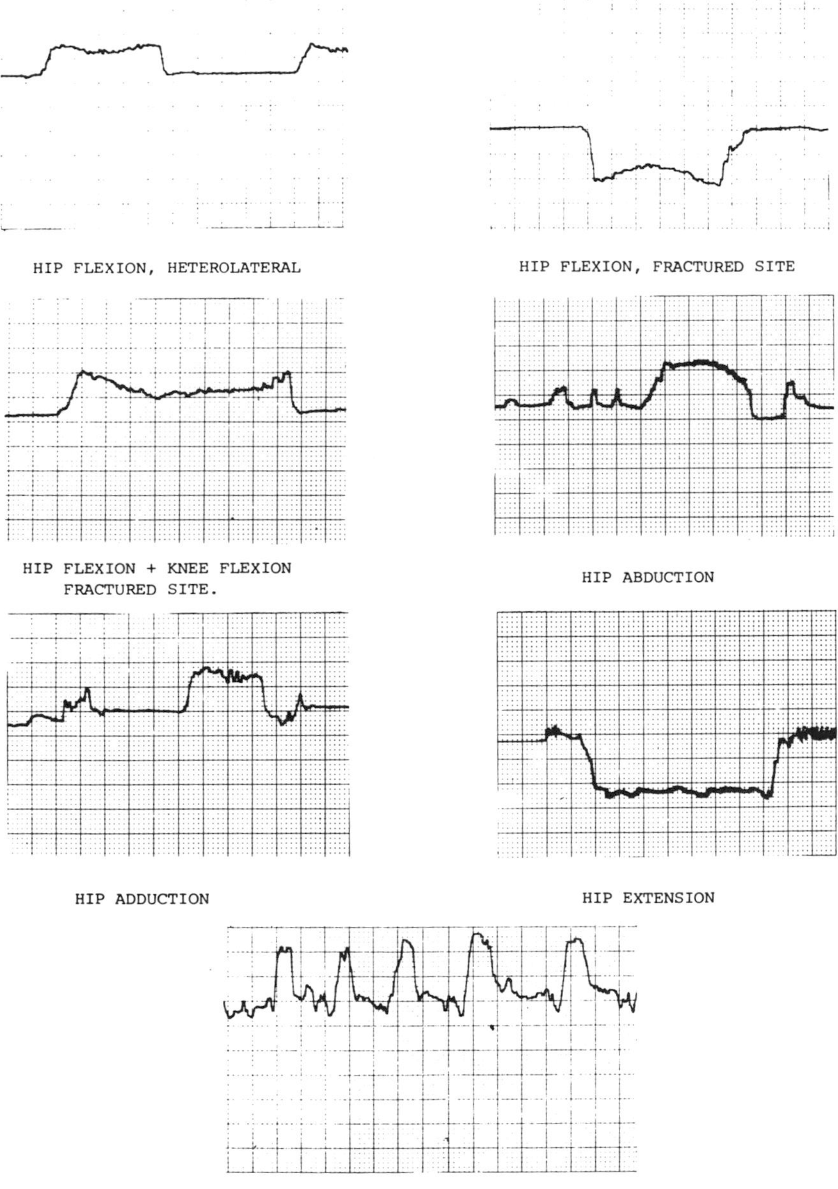

Fig. 5.3. Recording during rehabilitation of a fracture of the upper femur.

DISCUSSION

Biocompatibility of the transducer

The biocompatibility of the insulation materials was investigated using animals. Cylinders were implanted in the subcutaneous tissue of rats. In all cases we found giant cells (foreign body reaction), but never any malignant transformation (Burny *et al.* 1978). Histological investigations were performed on 15 clinical cases after retrieval of the implant. The same reactions were observed, without clinical effects (Fig. 5.4).

Fig. 5.4. Appearance of the material at the time of retrieval.

Insulation of the transducer

One of the major problems is electrical insulation. The insulation was investigated *in vitro* (Bourgois *et al.* 1976; Donkerwolcke *et al.* 1982). *In vivo* measurements of the insulation are presented in Fig. 5.5.

An insulation of over 100 MΩ, for six months, is obtained with strain gauge TML–PS3 bonded with Cyanolit and with Stycast insulation.

Effects of prestressing the implant

A possible prestressing effect could appear during surgery owing to the fitting of the implant to the bone. We measured prestressing during the implantation of upper femoral devices (Table 5.4). It does not seem to be significant (Burny *et al.* 1982).

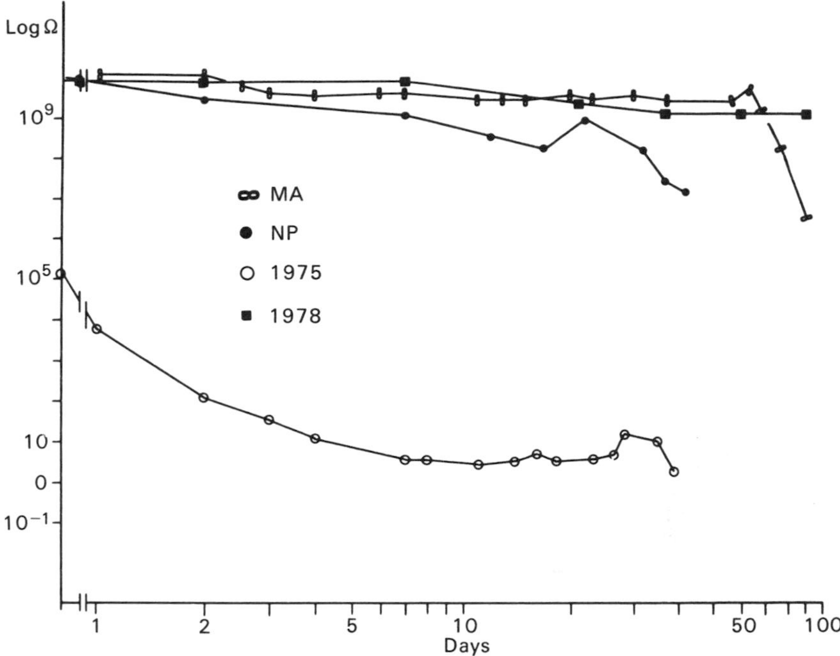

Fig. 5.5. *In vivo* measurement of insulation. MA—strain gauge on a Harrington rod; NP—strain gauge on the nail plates; 1975—values obtained in 1975; 1978—values obtained in 1978.

Table 5.4. *Prestressing after implantation*

Fracture	Nail plate	Location of SG	Prestressing (μs)
Pertrochanteric	experimental	junction	+90
Subtrochanteric	MacLaughlin 9	screw 1-3	−100
Subtrochanteric	MacLaughlin 9	screw 1-3	+36
Pertrochanteric	MacLaughlin 6	junction	+120
Subtrochanteric	experimental	junction	−176
Pertrochanteric	MacLaughlin 6	junction	−130
Subtrochanteric	MacLaughlin 12	screw 3-3	−368
Pertrochanteric	MacLaughlin 6	screw 1-3	−27

The transmission lead

Signal transmission requires transcutaneous leads. The reactions related to the lead are presented in Table 5.5.

Table 5.5. *Reactions related to the lead*

Type of reaction	frequency
No reaction	44
Staph. albus	17
Staph. aureus (alone or in association)	13
Other	8
Unknown	18
Total	100

The lead is kept in place until necessary (recording of low level of deformation) or until the appearance of pathological bacteria. The mean time is 36 days.

A comparison between wound infection and reactions around the lead was carried out. In all infected cases, bacteria appeared on the lead (*Staph. aureus*: four cases; other bacteria: two cases).

The infection rate of 6 per cent is not statistically different from the infection rate of a series of regular nail plates (5 per cent; Panda and Burny 1976).

It seems that contamination of the lead tract is secondary to wound infection.

Walking and weight bearing

Walking, with or without weight bearing, is allowed according to the measured deformation. In some cases, delayed weight bearing was imposed because of the risk of implant deformation or of fatigue failure.

We have information on walking for 77 patients, and on total weight bearing for 57 patients (Table 5.6). Twelve patients died before gait rehabilitation.

The major advantage of the technique is to decrease the risk of failure.

Table 5.6. *Walking*

Delay (days)	Walking (no. patients)	Total weight bearing (no. patients)
0–10	15	10
11–20	29	12
21–30	12	11
31–40	2	4
41–50	9	7
51–60	2	1
over 60	8	12

The characteristics of failure

The failure of an implant is not clearly defined in orthopaedic surgery. To be able to analyse and to compare statistics the classification given in Burny *et al.* (1985) has been proposed.

CONCLUSION

Using *in vivo* monitoring of implants we were able significantly to reduce the rate of failure of nail plates. With good nursing, no major complications occur at the site of penetration of the lead.

REFERENCES

Bourgois, R., Burny, F., Lorini, G., and Hermanne, A. (1976). Problèmes posés par l'utilisation des jauges de contrainte in vivo. *Acta orthopaed. belg.* **42**, Suppl. 1, 47–51.

Burny, F., Andrianne, Y., and Quintin, J. (1985). General model of failure of implant. In press.

—, Bourgois, R., and Aubriot, J. H. (1972). Causes des défaillances du matériel d'ostéosynthèse. In Proceedings of the 12th Congress of SICOT, Tel Aviv. International Congress Series No. 291, pp. 992–4.

—, Donkerwolcke, M., and Bourgois, R. (1982). Mesures per-opératoires de la précontrainte imposée aux implants d'Ostéosynthèse. *Acta orthopaed. belg.* **46**, Suppl. 1, 119–27.

—, —, Potvliege, P., and Magerat, M. (1978). *Biocompatibilité du matériel d'isolement utilisé pour la protection des jauges de contrainte in vivo.* Laboratoire de Chirurgie Expérimentale, ULB.

Decoulx, P. and Lavarde, G. (1969). Les fractures trochantériennes. In *71ème Congrès Français de Chirurgie*. Association Française de Chirurgie, Paris.

Donkerwolcke, M., Burny, F., Bourgois, R., and Quintin, J. (1982). Technique d'isolement des jauges de contraintes utilisées in vivo. Résultats d'études in vitro et résultats cliniques. *Acta orthopaed. belg.* **46**, Suppl. 1, 107–18.

Moulart, F. (1974–75). Réducation des fractures hautes du fémur. In *Mesure de la déformation du matériel d'ostéosynthèse*. Mémoire de Licence en Kinésithérapie, ULB.

Panda, M. (1974). Fractures sous-trochantériennes et trochantéro-diaphysaires. Thèse de Licence en Chirurgie Générale, Université de Kinshasa, Université Libre de Bruxelles.

— and Burny, F. (1976). Traitment des fractures sous-trochantériennes et trochantéro-diaphysaires. *Acta orthopaed. belg.* **42**, 429–44.

Ramadier, J. O., Duparc, J., Rougemont, D., and De Ferrari, D. (1956). Le traitment chirurgical des fractures trochantériennes et juxta-trochantériennes. *Revue Chir. orthopaed.* **42**, 759–82.

6 The *In Vivo* Measurement of Fracture Movement

K. E. TANNER

INTRODUCTION

Even before the time of Hippocrates (420 BC) the main principle of fracture treatment had been to reduce the movement of fractures to a greater or lesser extent. More recently, Yamagishi and Yoshimura (1955) theorized that shear forces at the fracture site produced pseudoarthroses, and various other writers, including Sarmiento (1970) and Bradley *et al.* (1979), have suggested that intermittent compression leads to improved fracture healing.

Lippert and Hirsch (1974) used transcutaneous bone pins supporting targets to measure photogrammetrically the movement of the bone segments in patients with tibial fractures at various stages of healing treated with a variety of types of plaster of Paris. Seligson *et al.* (1981) used strain-gauged hemicircles to measure the shear and bending movements at the fracture site in cadaveric tibiae fixed with an external fixator, or a Kuntscher nail, and subjected to cyclic bending moments.

In order to investigate the role of movement in fracture healing, it was decided to measure the six possible motions—three linear displacements and three angulations—at the fracture site. Those patients whose fractures were treated with an 'Oxford' fixator already had a firm mechanical connection between their bones and the outside world, and it was convenient to make use of this in measuring the fracture movements.

METHOD

A three-dimensional movement transducer based on three lateral effect photodetectors was built. These detectors (United Detector Technology, Inc.) are square centimetres of p-i-n diode, the planes of the diode being parallel to the surface of the detector. The n-base is metal and earthed; the p area, which is illuminated, has four contacts, one along the length of each side. From the currents at the four contacts the position of a spot of light in two dimensions can be found using one of the linearizing algorithms devised by Woltring (1975). Three of these detectors ((A), Figs. 6.1 and 6.2) were mounted orthogonally on the end of a bar (B) which is clamped to one of the Schanz pins used in the Oxford External Fixator. Three collimated infra-red light-emitting diodes (D)

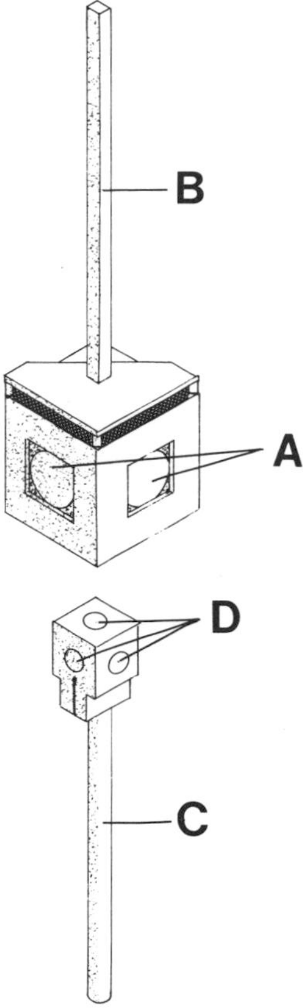

Fig. 6.1. Exploded view of the transducer.

are mounted orthogonally on another rod (C) which is clamped to another Schanz pin on the opposite side of the fracture. The clamps can be seen in Fig. 6.2. The two parts of the transducer are free to move relative to each other, and there is no load transmitted between them. The maximum movement is 1.5 mm in any direction from the initial set-up position. The outputs from the detectors are fed into current-to-voltage amplifiers and then into the A/D converter of a DEC PDP-11/34 computer. This then computes the displacements on each detector of its spot of light, and from these it calculates the relative movement of the two parts of the transducer.

When a fractured limb stabilized by an external fixator is loaded, the fixator deforms, producing movement at the fracture, and also between the two parts of the transducer. Almost all of the deformation of the fixator-bone system is due to the bending of the pin (Evans *et al.* 1979). Thus, if the pins are deforming elastically, rather than plastically, the deformation of the pins at the pin–bone

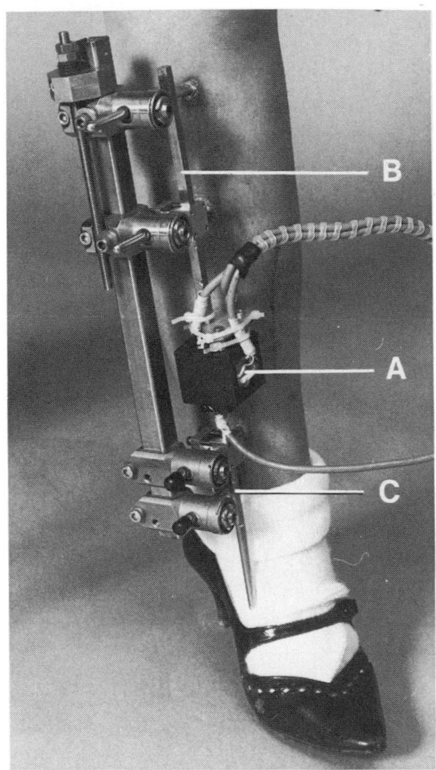

Fig. 6.2. The transducer in use on a patient.

interface and at the transducer clamps may be calculated using standard elasticity theory. Plastic deformation can be ruled out, since a load of 1.5 times body weight needs to be applied to one end of an average length (50 mm) double encastré pin to produce plastic deformation. If it is assumed that both the support rods of the transducer and the bones are rigid, the relationship between the six transducer movements and the six fracture site movements may be calculated. It is necessary to know the angle of the pins relative to the fixator support bar, the distances of the transducer and the bone from the fixator support bar, and the distances from the pins to the fracture line and to the centre of the transducer.

Ideally the patients are tested fortnightly, although sometimes this is not possible. The tests described by Evans and colleagues in Chapter 4 are performed: straight leg raise, longitudinal tibial loading, weight transfer, and in some cases walking. Where appropriate the forces applied to the floor by the patient's foot are also measured.

RESULTS

The results of the straight leg raise will be described. This test involves the assistant raising the patient's leg by the heel with the patient sitting relaxed on an examination couch. After about one second the patient is then asked to hold the leg in the same position for a further second or so and then to lower the leg to the couch, and relax. The tests results in the fracture being subjected to a bending moment and shear force in one direction, followed by a bending moment and shear force in the opposite direction. Figure 6.3 shows the total

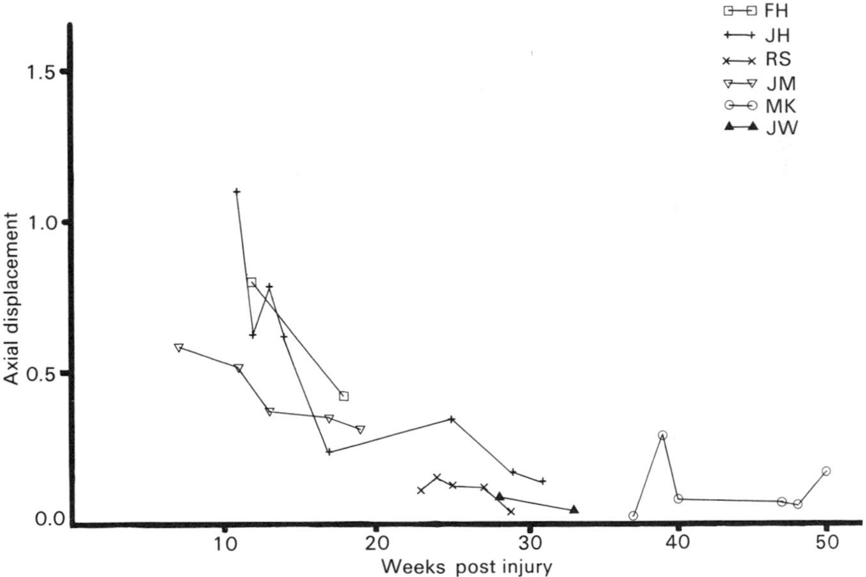

Fig. 6.3. The total movement at the fracture site versus time.

axial displacement, for various patients, between the positive and negative bending moments. In some patients the axial displacement drops swiftly in a hyperbolic type of curve, whereas in patients M.K. and R.S. this reduction is not present. The patient M.K., whose fixator was applied at 36 weeks for atrophic non-union and previous infection, had the fixator removed at 50 weeks, due to recurrence of the infection; three months later the patient was still in plaster of Paris, with further bone grafting planned. The patient R.S. was only included in this trial 23 weeks post-injury and fixation; it may be hypothesized that his fracture was already consolidated.

DISCUSSION AND CONCLUSIONS

This system is capable of measuring to an accuracy of approximately 0.03 mm. However, the main source of error is the scatter between two runs of the same test. Each of the points on the graph is the average of two readings, and the maximum difference between these readings was 0.2 mm. The movement at the fracture site is controlled by the forces being applied to the fixator/bone system, including the muscle forces, which are appreciable and also variable.

This is an acceptable method of measuring fracture healing. The tests take approximately 15 minutes, including the time required to apply the transducer to the patient. The position of the transducer does affect the calculations, but it was clamped to the same points on the fixator at each visit, and was checked by measurement. This system does not use X-rays, and shows the increasing stiffness, or otherwise, of the fracture with time. The increasing stiffness is apparent before the callus begins to calcify and become visible on radiographs.

REFERENCES

Bradley, G. W., McKenna, G. B., Dunn, H. K., Daniels, A. U., and Statton, W. O. (1979). Effects of flexural rigidity of plates on bone healing. *J. Bone Jt Surg.* **61A**, 866–72.

Evans, M., Kenwright, J., and Tanner, K. E. (1979). Analysis of a single-sided external fracture fixator. *Engng Med.* **8**, 133–7.

Hippocrates (420 BC). *Hippocratic writings* (ed. G. E. R. Lloyd). Pelican Classics, London (1978).

Lippert, F. G. and Hirsch, C. (1974). The three-dimensional measurement of tibia fracture motion by photogrammetry. *Clin. Orthopaed. related Res.* **105**, 130–43.

Sarmiento, A. (1970). A functional below-the-knee brace for tibial fractures—a report on its use in 135 cases. *J. Bone Jt Surg.* **52A**, 295–311.

Seligson, D., Powers, G., O'Connell, P., and Pope, M. H. (1981). Measurement of fracture gap motion in external fixation. *J. Trauma* **21**, 798–801.

Woltring, H. J. (1975). Single and dual-axis lateral photodetectors of rectangular shape. *Inst. elect. Electron. Engrs Trans. electron Devices* **22**, 581–90.

Yamagishi, M. and Yoshimura, Y. (1955). The biomechanics of fracture healing. *J. Bone Jt Surg.* **37A**, 1035–68.

7 The Reduction in Cortical Stress Levels Around Transcutaneous Fixation Pins

M. T. MANLEY, L. HURST, AND R. DEE

INTRODUCTION

Loosening of transcutaneous pins in external fixation of fractures remains a significant problem with the method. Various factors such as thermal necrosis of bone during pin insertion, infection and pressure necrosis due to high levels of stress in the cortex have been implicated in the loosening process. As yet, the primary cause of loosening remains to be established.

The intent of this study was to investigate the stress distribution in cortical bone around transcutaneous pins, and thereafter to identify pin designs which could allow a reduction in cortical stress levels *in vivo*. Since stresses in the cortex cannot be measured directly, a photoelastic technique was employed so that interface stresses close to pins of various designs could be compared.

MATERIALS AND METHODS

Three separate experiments were performed in this study. In the first, smooth stainless steel pins (316L) 4, 5, 6, and 7 mm diameter were inserted into 2D photoelastic epoxy resin (PSM5) models of cortical bone. Models were mounted sequentially in a transmission polariscope with the long pin parallel to the light beam. A load of 125N was applied to the pin and an interferogram recorded. The model was then rotated so that the pin was positioned at 45° to the light beam and a second, oblique, interferogram recorded. The fringe order and fringe distribution in these normal and oblique views allowed principal stresses to be calculated at any point in the model.

In the second experiment, smooth stainless steel pins of various diameters were coated with circumferential layers of polymeric materials of various thickness to produce composite pins of 4, 5, 6, and 7 mm outside diameter. Coatings were applied to the pins using electro-spray and thermal fusion techniques. The modulus and the hardness of the applied thermoplastics were not modified by the single fusion process. The interface materials investigated were silicon elastomer, nylon 11, halar ECTFE, UHMWPE, and acrylic. Testing methods for the coated pins were identifical to those used in the first experiment.

The third experiment was designed to investigate the effects of thread profile design upon cortical bone interface stress. Enlarged models (25X) of longitudinal sections of different pin designs were fabricated from zinc plate. Mating bone models were fabricated from CR30 epoxy resin. The modulus ratio of zinc to CR39 is close to that between stainless steel and cortical bone. Testing methods were consistent with those described for the smooth pins.

RESULTS

As expected the comparison of interface stresses around stainless steel bone pins of different diameters showed a decrease in interface stress at all points in the model with increase in pin diameter. However, it was found that the application of a low modulus interface coating further reduced stress levels.

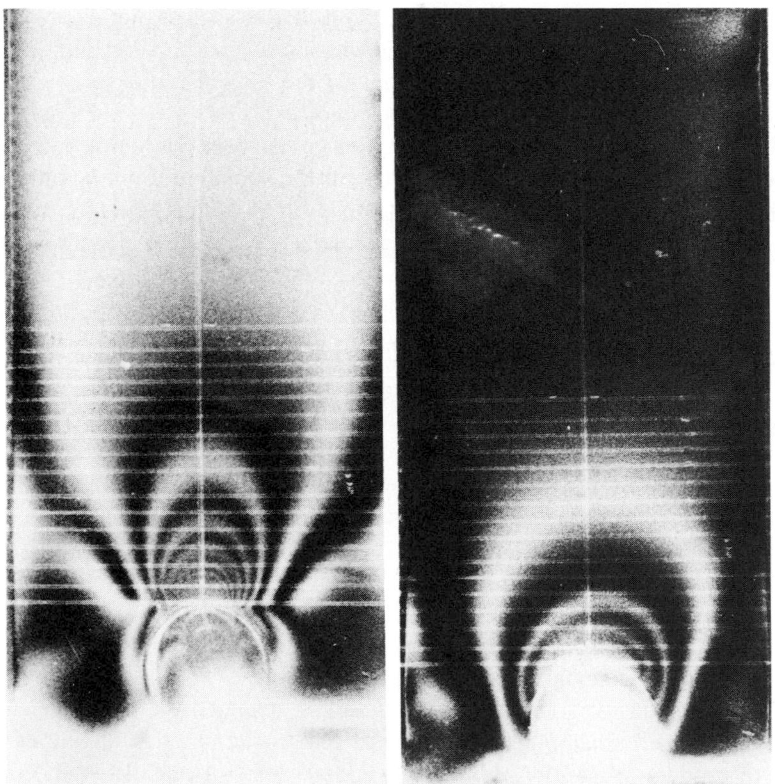

Fig. 7.1. Normal Incidence Interferograms of bone models loaded through a 6 mm diameter stainless steel pin (a) and a 4 mm diameter pin with 1 mm thick coating of silicon elastomer (b). The lesser number of fringes associated with the coated pin indicates that a lesser compressive stress exists at the pin–bone interface.

Normal incidence interferograms of a 6 mm diameter 316L pin and a 4 mm diameter 316L pin with 1 mm silicon elastomer coating are shown in Fig. 7.1A and B. The stainless steel specimen produced interference fringes which radiate deeper into the model when compared to the coated specimen of the same diameter. The larger number of fringes which surround the uncoated pin are indicative of a higher stress level at the pin–bone model interface. It is thus apparent that the presence of the silicon elastomer layer has reduced stress levels close to the pin, and that this reduction has been achieved without a change in the overall pin diameter.

The comparison of the effect of four different types of interface coating materials upon fringe order (and hence interface stress) is shown in Fig. 7.2. Each of these pins were 6 mm outside diameter with a coating thickness of 1 mm on the pin radius. All of the interface coatings caused some reduction in interface stress when compared to a 6 mm diameter stainless steel pin. However, it is apparent that silicon elastomer was the most effective at reducing interface stress. A fringe order of 5 at the pin surface for the silicon pins can be compared to 10 for bone cement, 11 for halar and 14 for nylon.

Figure 7.3 shows the effects of different pin thread profiles on interface stress. It is noticeable that all of the 'traditional' threads (A–E) produce similar stress concentration factors in the bone models. By comparison thread F, which presents the largest flat face to the applied compressive load, succeeds in reducing interface stress levels close to the pin.

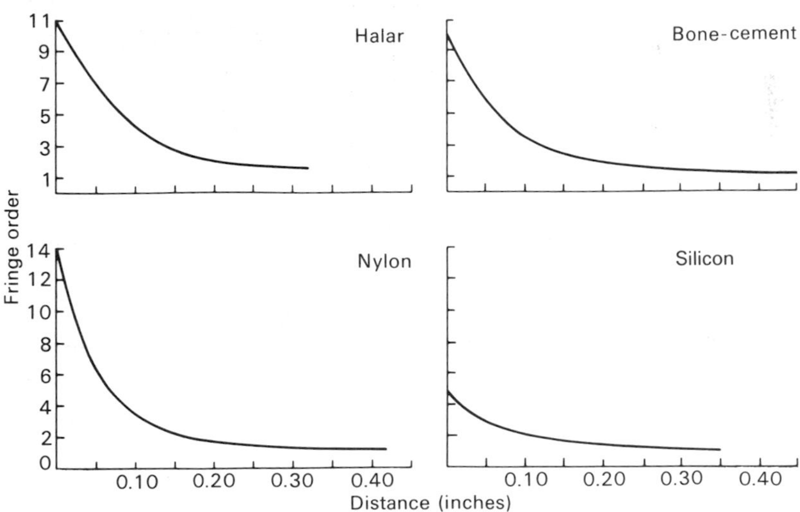

Fig. 7.2. The fringe distribution in models loaded through pins coated with different materials. The smaller number of fringes associated with the silicon elastomer coating indicates that it is most effective in reducing compressive stress close to the pin.

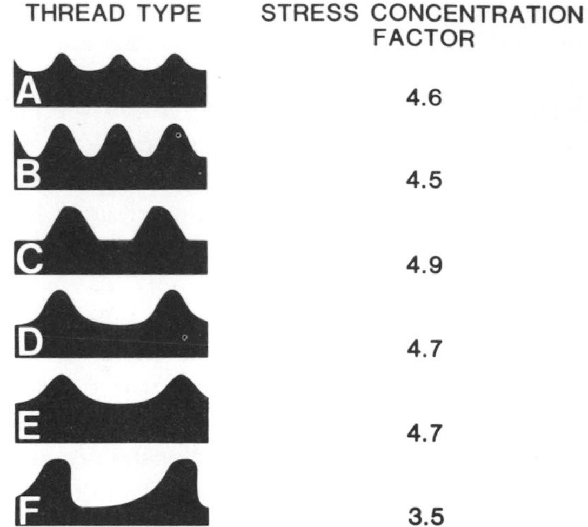

Fig. 7.3. The maximum stress concentration factors associated with bone screws of different thread types.

CONCLUSIONS

This study indicates that the magnitude of compressive stresses around transcutaneous pins can be reduced by maximizing pin diameter, by coating the pin with a low modulus interface material and by optimizing pin thread profiles. A full range of biocompatible polymeric materials are now being investigated as interface coatings. If a successful bond strength can be achieved between coating and pin this study will be the basis for a new design of transcutaneous fixation pins.

Acknowledgement

This project was supported by a grant from Zimmer USA.

8 Functional Index: A New Objective Assessment Applicable to Wrist Fractures

M. L. PORTER, I. STOCKLEY, AND W. K. PURVES

INTRODUCTION

The development of objective techniques for the accurate assessment of wrist function is required for both clinical and research purposes. Before this study, no satisfactory method of functional assessment existed. Methods of functional evaluation used in previous studies are mainly subjective in nature and vague in their criteria of categorization (Hollingsworth and Morris 1976; Cassebaum 1950). A fair result by one observer, for example, may be regarded as poor by another. A semi-objective assessment was described by Gartland and Werley (1951), who used a points system based on a combination of subjective and objective assessments. As a result of these varied and inaccurate methods of evaluation, there are conflicting reports concerning the functional impairment that follows a fracture to the distal radius (Bacon and Kurtzke 1953; Lidstrom 1959; Cooney *et al.* 1979). This chapter outlines a new objective method of assessment using specially designed apparatus.

METHODS

Eighty-nine patients with fractures of the distal radius were seen in the Orthopaedic Department, Manchester Royal Infirmary, in the six-month period from August 1981 to February 1982. A functional evaluation was carried out at a review clinic six months after initial injury. The group consisted of 71 females and 28 males, whose average age was 60 years (range 17–90 years). The objective assessment involved testing both the fractured and uninjured wrist using three separate pieces of equipment.

The torque dynanometer, shown in Fig. 8.1, was constructed to record wrist rotation while acting against a resistive torque. The device consists of two similar ratchet mechanisms connected about a central shaft, one applying resistance to clockwise rotation and the other applying resistance to anticlockwise rotation. A hand-grip terminates the shaft at one end. A measurement scale fixed to the other end quantifies the amount of rotation achieved by the patient

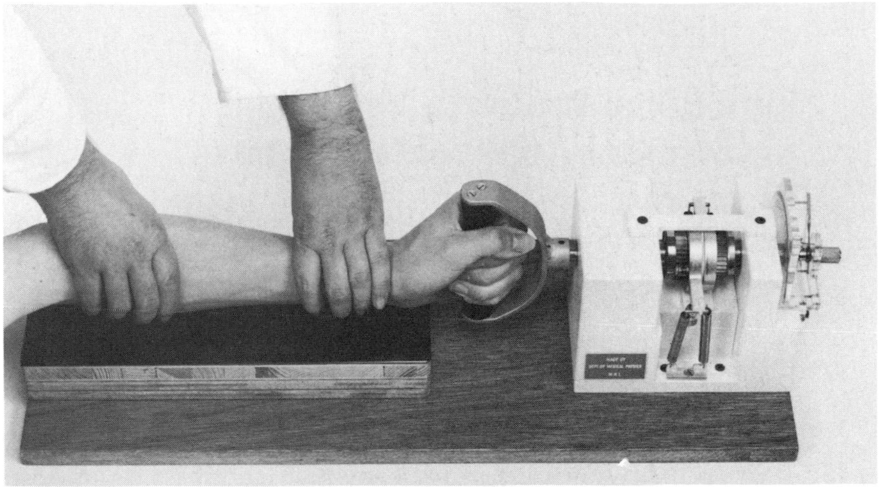

Fig. 8.1. Dynanometer. (Photo: Porter/Stockley.)

in degrees. The shaft rotates the sweep needle which in turn rotates one of the peak needles giving a maximum reading.

The patient was positioned at a suitable height in front of the device. The forearm was placed in the longitudinal axis on the instrument platform, with fingers grasping the handgrip. One of the examiner's hands was placed around the patient's upper forearm and the elbow located in the neutral position. This was established by ensuring that the medial and lateral humeral epicondyles were in a plane perpendicular to the instrument platform. The other hand of the examiner was placed loosely around the patient's wrist to stabilize but not restrict forearm movement. The patient was then asked to rotate the handle but told to relax as soon as lateral elbow movement was detected by the examiner. This end point signified shoulder joint rotation. The patient was allowed as many practice rotations as necessary until feeling conversant with the test requirements (usually two or three turns). Both clockwise and anti-clockwise movements were tested and the readings noted on the peak needles. These two scores were then summated to give a single reading. The test was then repeated on the contralateral wrist and the results recorded. A torque index (F_T) was then calculated as follows:

$$\text{Torque index} = \frac{\text{Score fractured wrist}}{\text{Score uninjured wrist}} \times 100\%.$$

Grip strength was measured using a standard vigorometer shown in Fig. 8.2. This device is supplied with three squeeze balls, and for the purposes of this study the medium sized ball was used. The ball was placed in the palm of the hand and the patient told to squeeze. The pressure was displayed on the measurement gauge.

Fig. 8.2. Vigorometer. (Photo: Porter/Stockley.)

Practice sessions were allowed before readings were taken for both hands. A grip index (F_G) was similarly calculated:

$$\text{Grip Index} = \frac{\text{Score fractured wrist}}{\text{Score uninjured wrist}} \times 100\%.$$

The goniometer shown in Fig. 8.3 was used to record the total active range of movement for each wrist. This instrument consists of a flat base with parallel grid lines drawn on the proximal half of its surface. The distal surface is covered with two side-walls, an end-wall, and roof made of perspex. The fixed measurement scale on the roof of the instrument is used to measure wrist flexion, extension, radial deviation, and ulnar deviation. A metal pin passes vertically through the centre of the scale. This pin can be raised and lowered by releasing the screw on the side of the pin bearing. The pin is used to centralize the wrist before recordings are taken. In addition a perspex ruler is free to rotate around the zero axis of the protractor scale. This is used to record wrist movement.

A mobile measurement scale is mounted on the end of the instrument and is used to measure wrist supination and pronation. Its position over the face of the perspex is adjusted by means of the three twist screws positioned at the angles of the inverted V-shaped arm of the protractor scale.

The patient's forearm was positioned prone, parallel to the grid lines on the instrument base. The metal pin was lowered and placed over the centre of the

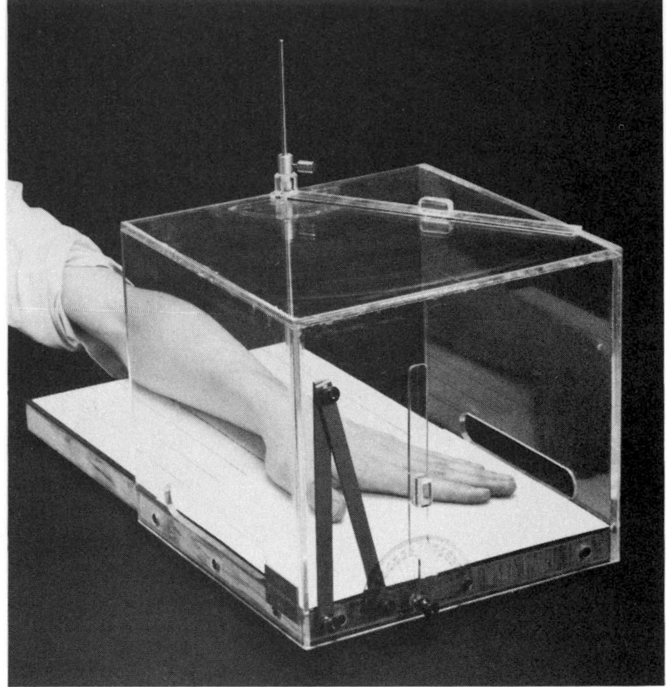

Fig. 8.3. Goniometer. (Photo: Porter/Stockley.)

wrist. This ensured that the wrist was centred over the axis of the protractor scale. The pin was then moved upwards out of the test field. The examiner's hands were positioned over the patient's forearm to abolish forearm movement but not restrict wrist motion. The patient was told to move the hand sideways into first ulnar and then radial deviation. At the same time the fingers were extended and approximated to avoid movement at the metacarpophalangeal joints. At the extremes of movement the perspex ruler was rotated and aligned with the longitudinal axis of the middle finger. Readings were then obtained off the protractor scale.

The forearm was then rotated through 90° with the radial side uppermost. The pin was again lowered and the forearm positioned parallel to the grid lines and with the pin overlying the styloid process of the radius. The fist was clenched and the examiners hands were again placed over the patient's forearm. The patient was then instructed to extend and then flex the wrist.

At the extremes of movement the perspex ruler was rotated and aligned with the V-shaped skin crease between the proximal thumb phalanx and second metacarpal bone. Readings were then obtained from the protractor scale.

Finally, the adjustable vertical scale was used to measure pronation and supination. The forearm was positioned parallel to the grid lines, elbow flexed and fingers extended in the neutral position. The examiner ensured that the medial and lateral humeral epicondyles were in a plane perpendicular to the instrument base—thus preventing shoulder rotation. The vertical protractor scale was positioned by means of the adjustable screws so that its axis was centred over the end of the middle finger. The patient was told to rotate the forearm clockwise and then anticlockwise to give supination and pronation readings. These were taken at the extreme of movement. The perspex ruler was aligned with the tip of the thumb and the scale readings recorded.

The six readings: ulnar deviation, radial deviation, flexion, extension, supination, and pronation were then summated to give a single score for each wrist. A range of movement index (F_R) was then calculated:

$$F_R = \frac{\text{Total score fractured wrist}}{\text{Total score uninjured wrist}} \times 100\%.$$

Each index, F_T, F_G, and F_R is used individually for research purposes. The overall functional index (*FI*) is calculated as follows:

$$FI = \frac{F_T + F_G + F_R}{3}\ \%.$$

Other methods of assessment were made as follows:

1. By the patient into: Good
   Fair
   Bad
2. By the authors into: Good
   Fair
   Bad
3. Gartland's points system:

	Points
Excellent	0–2
Good	3–8
Fair	9–20
Bad	21+

RESULTS

A computerized statistical analysis has been used to resolve several issues regarding functional index. First, do the functional indices and its components correlate with other assessments of function? Secondly, do the indices, F_T, F_G, and F_R measure the same aspect of function? Does hand dexterity bias the functional index? Finally, can the objective results be expressed in a more accurate way?

Correlation with other assessments of function

The functional indices (F_T, F_G, F_R, and FI) were individually compared with the subjective assessments made by: (i) the patient; (ii) the authors; (iii) Gartland's point system. The median scores are illustrated in Tables 8.1–8.3 respectively.

Table 8.1. *Comparison of functional indices with subjective assessment made by patient (median scores). Kruskal Wallis one way test*

	Good	Fair	Bad	χ^2	*p* value
F_T	96	90	88	10.89	$p = 0.004$
F_G	51	46	17	9.75	$p = 0.008$
F_R	93	85	68	17.78	$p < 0.001$
FI	86	73	62	15.16	$p < 0.001$

Table 8.2. *Comparison of function indices with subjective assessment made by authors (median scores). Kruskal Wallis one-way test*

	Good	Fair	Bad	χ^2	*p* value
F_T	96	90	88	13.30	$p = 0.001$
F_G	60	46	04	16.63	$p < 0.001$
F_R	95	85	69	23.54	$p < 0.001$
FI	86	73	55	26.20	$p < 0.001$

Table 8.3. *Comparison of functional indices with Gartlands point score*

	Kendall correlation coefficient	*p* value
F_T	0.2615	$p = 0.001$
F_G	0.2797	$p = 0.001$
F_R	0.3951	$p = 0.001$
FI	0.4101	$p = 0.001$

In each case there are statistically significant differences between the groups. The trend is for an increase in index with subjective improvement in fucntion. This implies that each individual index is capable of measuring function.

Intercorrelation

F_T, F_G, and F_R were independently plotted against each other. The Kendall correlation coefficients for the three scatterplots were 0.2486, 0.2641, and 0.2471. These plots are shown in Figs. 8.4–8.6. If each index measured the same aspect of function then a good correlation between the indices would be expected.

On inspection of the scatterplots the intercorrelation is poor. This implies each index measures a different aspect of function.

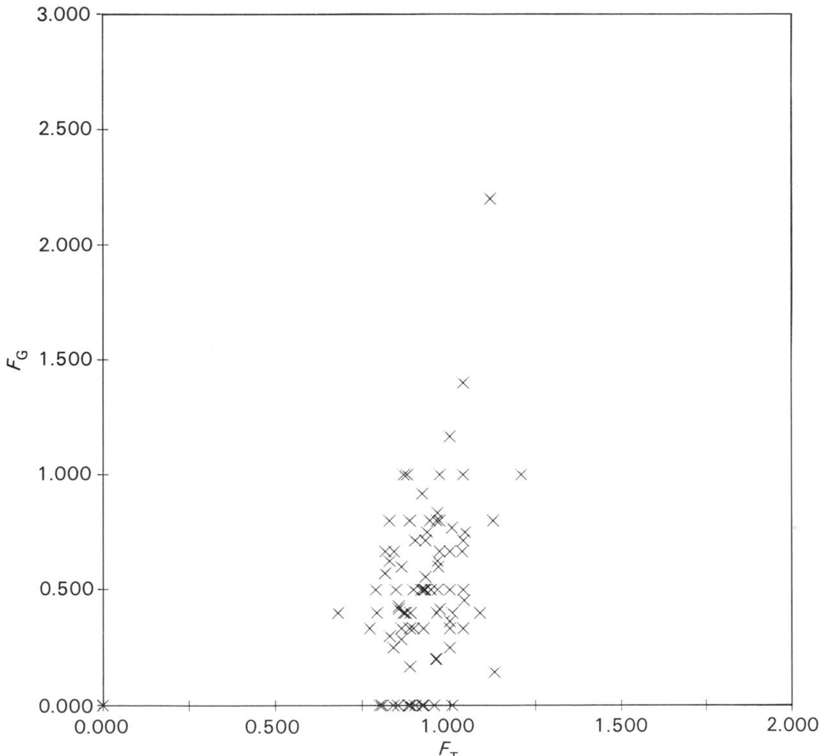

Fig. 8.4. F_G plotted against F_T. X score 2.

Hand dexterity

Eighty-one patients were right handed; 41 fractured their dominant hand and 40 fractured their non-dominant hand. A comparison was made between the absolute scores of the non-fractured wrists of these two groups. In all cases there was no significant difference between the scores of these two groups (Mann Whitney U test).

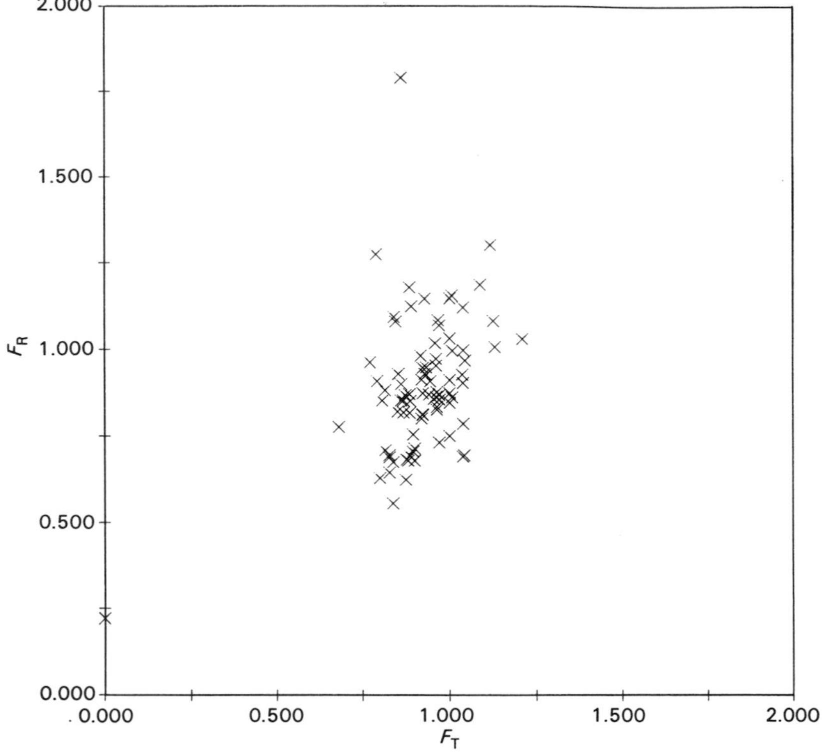

Fig. 8.5. F_R plotted against F_T. X score 2.

Principal component analysis

A principal component analysis was used to establish whether the objective information obtained from the three tests could be recombined or used in a more meaningful way. Principal components were identified but these generally reflected the same information obtained from the functional indices. In addition, these components are mathematically more complicated and less sensitive. The representation of functional index in its present form is more effective.

DISCUSSION

A wrist joint is required to perform under a variety of conditions if normal function is to be expected. An ankylosed or stiff wrist, for example, is no longer able to work dynamically, consequently function will be impaired.

Functional index, therefore, was designed to test the wrist under three different biomechanical conditions. The dynanometer imposes rotational torque conditions. The index obtained is the torque index, F_T.

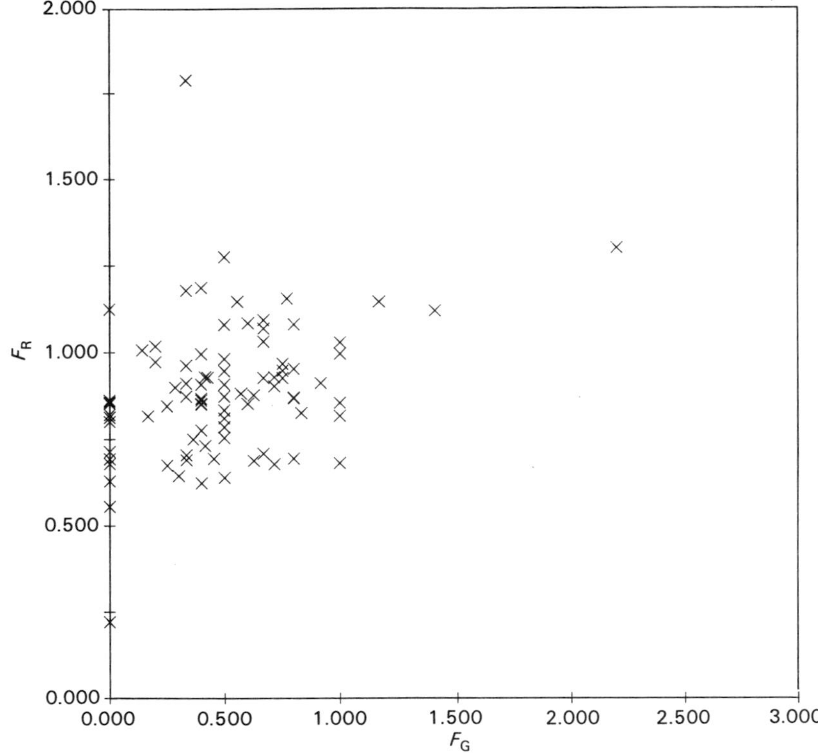

Fig. 8.6. F_R plotted against F_G. X score 2.

The vigorometer tests the ability of the wrist to transfer load from the flexor muscles of the forearm through the wrist to the finger joints. The index obtained is the grip index F_G.

The goniometer measures the performance of the wrist allowing active movement, without loading. The index obtained is the range of movement index F_R.

The overall functional index, *FI*, is an average of the above three indices.

All the tests, to some extent, are dependent on factors apart from intimate wrist joint function. Wrist torque, for example, is partly reliant on hard grip, muscle strength, and superior radio-ulnar movement.

These factors are of secondary importance, as in all tests a direct comparison is made with the non-injured wrist and the result expressed an an index. In addition, minor discrepancies in testing technique are tolerated as long as the forearm is correctly positioned on the test machine and both wrists are tested in the same way.

Apart from being a highly objective technique, functional index has several other attractions. The machinery is straightforward to construct and easy to use (by paramedical staff if desired). The average testing time is short (5–10 minutes). Above all, the result is quantifiable and is expressed usefully, e.g. FI = 80 per cent implies that the injured wrist has 80 per cent of the overall function of the contralateral wrist.

Functional index has both clinical and research applications. Clinically, it can be used to monitor the response to treatment following wrist fracture. A steadily improving index implies a response to treatment. Functional index is a useful research tool. Previous studies on wrist fractures have been difficult to compare as different subjective assessments have been used. The functional index can resolve this problem. Functional index has been introduced to a recent series of wrist fractures. Important factors influencing outcome have been identified which would not have been revealed using more inaccurate subjective methods of assessment. It is hoped, shortly, to publish the results of this study.

Acknowledgements

The authors wish to thank the staff of the Medical Physics Mechanical Workshop, Manchester Royal Infirmary, who built the equipment, Mr Markham and Mr Maltby, Consultant Orthopaedic Surgeons at the Manchester Royal Infirmary for allowing us to study their patients, Miss Linda Hunt for carrying out the statistical computer analysis, and Miss Wendy Dainty for typing the manuscript.

REFERENCES

Bacon, R. W. and Kurtzke, J. F. (1953). Colles' fracture. A study of two thousand cases from the New York State Workmen's Compensation Board. *J. Bone Jt Surg.* **35A**, 643–58.

Cassebaum, W. H. (1950). Colles' fractures. A study of end results. *J. Am. med. Ass.* **143**, 963–5.

Cooney, W. P., Linscheid, R. L., and Dobyns, J. H. (1979). External pin fixation for unstable Colles' fractures. *J. Bone Jt Surg.* **61A**, 840–5.

Gartland, J. R. and Werley, C. W. (1951). Evaluation of healed Colles' fractures. *J. Bone Jt Surg.* **33A**, 895–907.

Hollingsworth, R. and Morris, J. (1976). The importance of the ulnar side of the wrist in fractures of the distal end of the radius. *Injury* **7**, 263–6.

Lidstrom, A. (1959). Fractures of the distal radius. A clinical and statistical study of end results. *Acta orthopaed. scand.* Suppl. 41.

9 Some Comparative Properties of Splintage Materials

D. J. PRATT, E. S. POWELL, D. I. ROWLEY, S. H. NORRIS, AND T. DUCKWORTH

INTRODUCTION

Bandage-mounted plaster of Paris has been the principal splintage material in orthopaedics for a long time, its use being reported in 1798 by Mr Eton (Monro 1935). Before this time a number of other materials had been used including various resin-impregnated cloths. Recently the predominant use of plaster of Paris on cotton bandage has been challenged by a number of products termed 'second generation' casting materials. Most of these are variations of resin-impregnated bandages and it is important to measure their properties in order to define a role for each product.

MATERIALS

The materials chosen for study were selected from the range of current 'second-generation' casting materials and were all water-activated bandages. Gypsona (plaster of Paris) was also included as a standard.

Below is a list of products tested:

Gypsona	Smith & Nephew	Anhydrous calcium sulphate mounted on a cotton bandage
Crystona	Smith & Nephew	Alumino-silicate glass and acrylic acid polymer on cotton bandage
Zoroc	Johnson & Johnson	Plaster of Paris with resin to reinforce and waterproof
Baycast	Johnson & Johnson	Cotton bandage impregnated with a pre-polymer that forms cross-linkages in contact with water
Scotchcast	3M	Knitted fibreglass bandage with polyurethane resin
Scotchflex	3M	As above but the resin is mounted on a lighter more flexible mesh bandage

METHODS

Before selecting the methods for testing the products, the criteria for an ideal cast material had to be defined. Primarily the material should have a high strength-to-weight ratio, provide sufficient support for the limb and allow adequate radiological examination of the limb with the splint *in situ*. Other desirable attributes of the material should be that it be easy to work and mould, clean to apply, easy to remove, and be waterproof. It would be difficult to design a trial to assess the aesthetic qualities of the products. On the other hand it would seem reasonable to assume that the clinician's choice should be influenced by the physical properties of the various materials. Tests were thus devised to compare the products listed under control conditions and to compare the results obtained with those for plaster of Paris, the standard product used. The tests do not try to duplicate clinical conditions but they do provide useful comparative figures for selecting one of the products for a particular application.

The tests selected were as follows:

Three- and four-point bending

Rectangular slabs of material nominally 25 mm wide and between two and eight layers thick were prepared from standard bandages following the manufacturers instructions; these were rolled with a roller to ensure good lamination. All mechanical test samples were dried at room temperature for one week to ensure that all products had attained maximum strength. They were then loaded into an Instron testing machine in three- and four-point bending. The jig for three-point bending was made from 10 mm duralumin plate and the load-bearing elements consisted of 6 mm diameter steel pins. A span of 76.5 mm was used, being based on BS 2782 (1978), with the central support at mid-point. The central support was connected to the cross-head of the Instron and this was set to rise at 50 mm per minute. The jig for four-point bending had the same bottom half but the top half now consisted of two load-bearing elements separated by 25.5 mm and made of the same material as in the three-point bending jig. The same cross-head rise rate was also used for this test. Both tests were performed as the different loading geometries produce different bending moments along the test piece. In four-point bending a pure tension/compression situation pertains with no shear forces between the central two points which is not the case in three-point bending. It was felt that the different loading situations would provide extra information regarding the suitability of a product. Each product was tested to failure, the load at which this occurred being noted.

A load cell within the Instron testing machine enabled a graph of load against cross-head travel to be plotted directly.

Flexural modulus

A graph of the initial slopes of the three-point bending load deflection curves at each thickness plotted against $4bd^3/L^3$ was plotted (where b and d are test pieces width and thickness respectively and L is the span length). The gradient of the graph thus obtained is equal to the absolute flexural modulus. Specific flexural moduli were calculated by dividing the absolute value by the specific gravity of the product.

Radiolucency

A 7.5 cm thick block of tissue-equivalent beeswax was fashioned. On the top and bottom faces of this slab were placed sheets of the products from two to eight layers thick (i.e. 4–16 in total). To simulate a diagnostic radiograph, standard XG1 X-ray film with a fine screen was used at one metre from the X-ray head with the slab on top. An exposure of 70 KVp at 100 mA for 0.15 s was made and the density of the film under each thickness of material was measured, once developed, using a densitometer. At least ten readings were made for each product at each thickness and the film density plotted against the number of layers.

Fatigue studies

Twenty-five millimetre wide slabs six layers thick were fabricated from standard bandage. They were then placed into a jig which flexed the sample through a fixed displacement using a cam-follower system. The sample was positioned such that the load applied at maximum displacement of the cam-follower was 50 per cent of the maximum load capacity measured during the three-point bending tests. The jig was cycled at 1 Hz and the load in the cam-follower measured using a strain-gauged transducer. The fatigue life was taken to be when the load on the follower at full displacement fell to 50 per cent of its initial value. The number of cycles at which this point occurred was recorded by a counter on the cam.

Compression of cylinders

Six centimetre long cylinders, 2.5 cm in diameter and from two to eight layers thick, were produced from standard bandages on a cylindrical former. These were subjected to axial compressive loading, the maximum load recorded for each sample being noted.

RESULTS

Figs. 9.1(a) and (b) show the results for the three- and four-point bending tests respectively. The points plotted are the average of the five samples tested at each thickness with ± one standard deviation indicated by the vertical bars. With both bending geometries Scotchcast is the strongest product, the remainder being grouped together. In three-point bending Gypsona (plaster of Paris) is found to be the weakest. All the products can withstand larger loads in four-point bending, the graphs tend to have a sigmoid shape and Crystona is found to perform particularly well at larger thicknesses.

From Fig. 9.1(a) the number of layers required to support a load of 6 N was measured and recorded (Table 9.1). This shows that there is a wide range of values (0.7–6.5 layers). Only one product, Crystona, requires more layers than Gypsona; the fibreglass bandage products are particularly good.

Table 9.1. *The number of layers of each product required to support a load of 6 N, and their relative specific strengths*

Product	No. of layers to withstand 6 N	Specific strength
Gypsona	6	1.00
Crystona	6.5	0.57
Zoroc	5.3	1.35
Baycast	5.5	7.52
Scotchflex	1.1	14.34
Scotchcast	0.7	23.29

The flexural moduli of the products (Fig. 9.1(c)), shows that in absolute terms Zoroc is the stiffest, but when the specific gravity of the products is taken into account, Baycast is the stiffest. Crystona is the most flexible specifically but Scotchflex is the most flexible in absolute terms.

Figure 9.2 shows the product radiolucencies where Baycast is seen to be the most radiolucent. The remainder are in a group bounded by Scotchflex at the top and by Zoroc and Gypsona at the bottom. The lines are drawn by linear regression analysis and the numbers on the right are the coefficients of correlation. The lines are all significant to greater than 95 per cent. Although there is a wide range of values none of the materials should present a problem for radiographers.

The results of the fatigue studies (Fig. 9.3(a)) show that Gypsona, Crystona, and Zoroc all have fatigue lives of less than 100 cycles. The remainder of the products have much higher values ranging from Baycast with 22 861 to Scotchcast with 39 704. The low value for Gypsona is particularly interesting when its common use is considered.

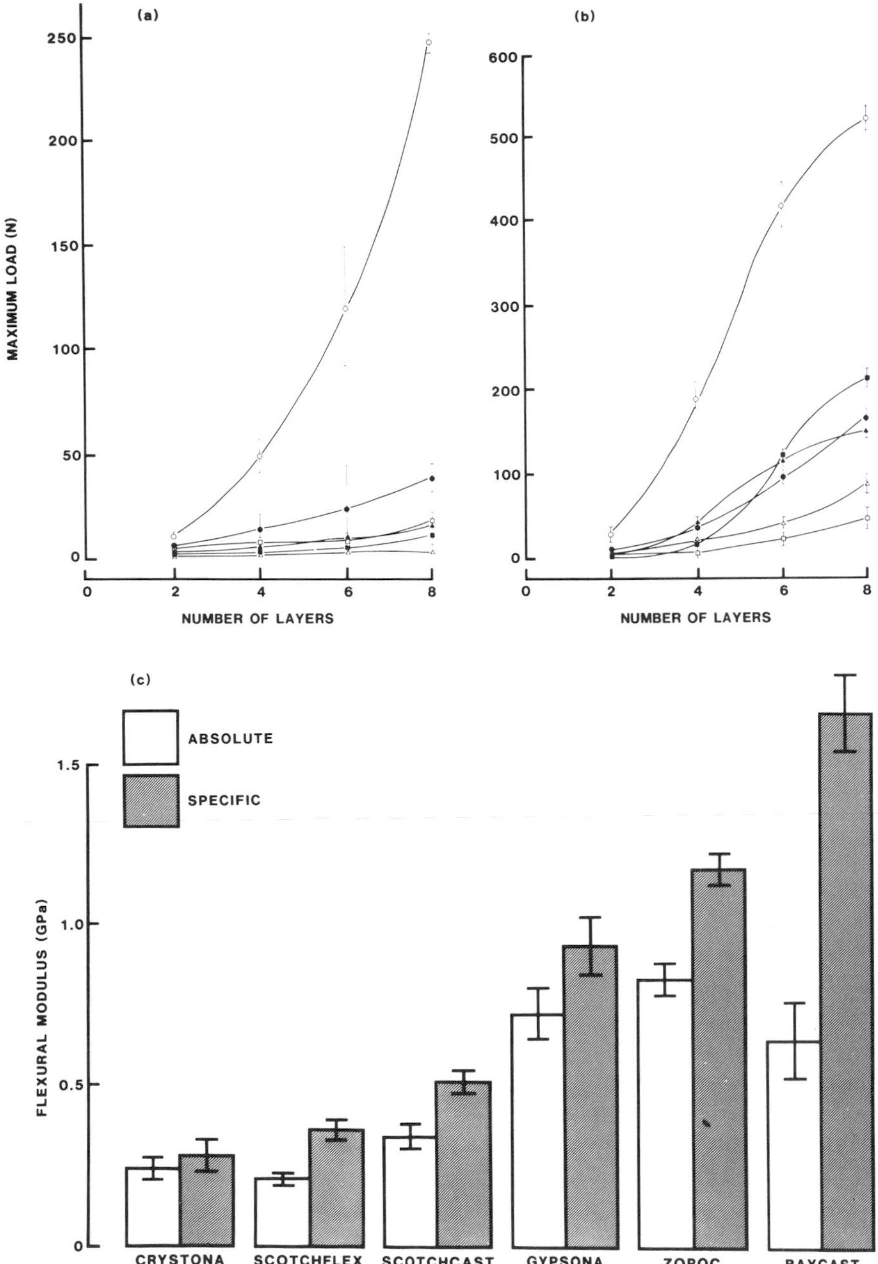

Fig. 9.1. Graphs of the results recorded in (a) three-point bending; (b) four-point bending; and (c) the flexural moduli obtained from the above (see text for details). The vertical bars represent ± one standard deviation about the mean. ○ = Scotchcast; ■ = Crystona; ▲ = Zoroc; ● = Scotchflex; △ = Gypsona; □ = Baycast.

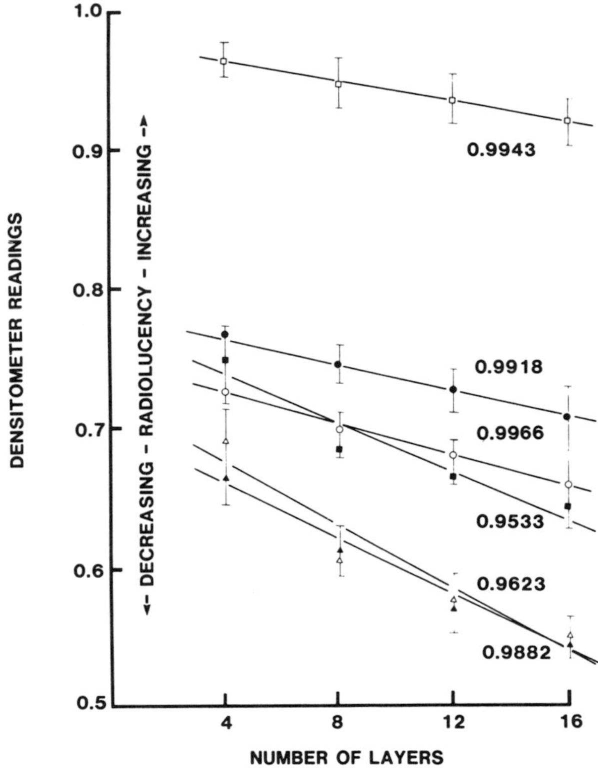

Fig. 9.2. Product radiolucency. On this scale a completely black radiograph has a value of 6 and completely clear one has a value of 0. ○ = Scotchcast; ■ = Crystona; ▲ = Zoroc; ● = Scotchflex; △ = Gypsona; □ = Baycast.

The results of the compression of cylinders are presented in Fig. 9.3(b). The loads recorded on this test are all very high, with Scotchcast the strongest, reaching over 12 kN for an eight-layer cylinder. Baycast was weakest, only reaching 1-2 kN at eight layers. Zoroc was the strongest of the group of products not including Scotchcast, reaching well over 4 kN for eight layers.

DISCUSSION

These results generally support those published by Gill and Bowker (1982). The present study includes a larger range of products but excludes Hexcelite, as it is not a bandage which sets or cures, but simply cools down. In the light of additional products now being marketed this study needs to be extended to include such products as Baycast Plus, Cellona, and Dynacast.

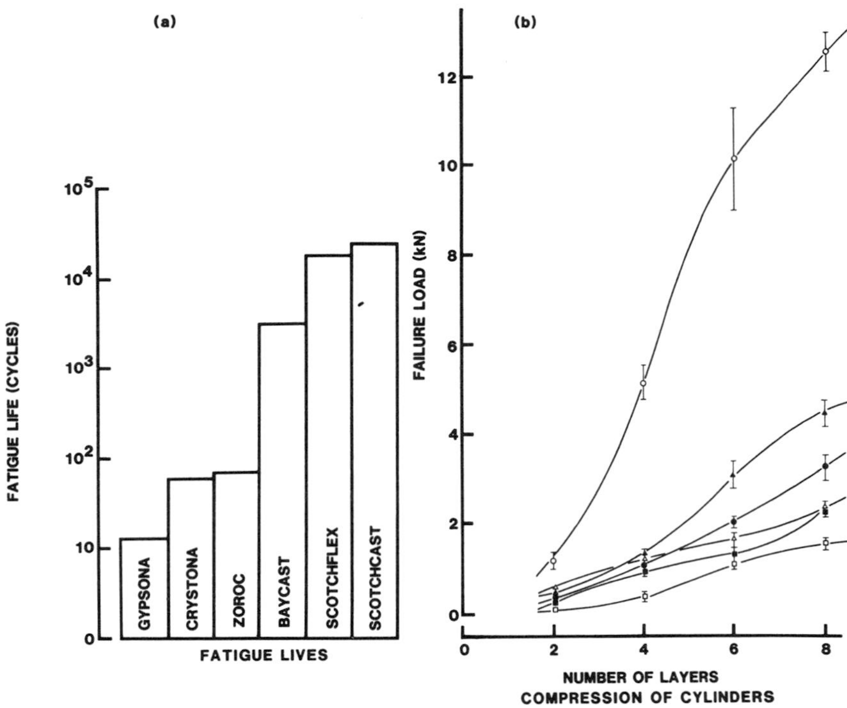

Fig. 9.3. Graphs of (a) the product fatigue lives (note: the ordinate is logarithmic) and (b) the axial compression of cylinders. ○ = Scotchast; ■ = Crystona; ▲ = Zoroc; ● = Scotchflex; △ = Gypsona; □ = Baycast.

When looking at the results presented, the strengths of the products, in whichever way measured, only give an indication of how the material might behave in clinical use. It is clear that any product which does not withstand clinically encountered loads is unsuitable for use in weight-bearing casts. However, the strength of the product is only part of the picture and the flexibility of the material has to be taken into account. High strain-rate loadings are the norm in weight-bearing casts and, for example, although Crystona has a lower specific strength than Gypsona the flexibility advantages of Crystona produce a more resilient cast. So an ideal material should have a good strength-to-weight ratio (SWR) and be flexible. Table 9.1 shows the specific strengths for the products tested. These SWRs were calculated by taking the load recorded during the three-point bending tests for six-layers slabs and dividing by the specific gravity. The value obtained was divided by that for Gypsona to give the specific strength. All the products except Crystona have a higher SWR than Gypsona, Scotchcast being the highest.

The fatigue life for an ideal material should be high, although an absolute value is difficult to assign. Of all the products tested it is perhaps fair to say that high values are preferable, since the number of stress reversals in a lower limb cast during a normal day is also high.

When considering the compression of cylinders, although Scotchcast is the strongest, the loads recorded for all the products would seem to satisfy the strength criteria for an ideal casting material. When considering lower loads on Fig. 9.1(a), it is seen that all of the products, except Crystona, require fewer layers of bandage to support a 6 N load, with the fibreglass products, Scotchcast and Scotchflex doing particularly well.

When all of these data are considered, it must be noted that the test conditions did not exactly simulate a clinical situation. The geometry of the cast and the presence of the limb within the cast will give non-ideal loading situations. These results only serve as guidelines to the clinical usefulness of the products.

One of the criteria for an ideal casting material is that it should allow adequate radiological examination whilst *in situ*. Baycast is by far the best product on this basis, although none of the products is worse than Gypsona, the accepted standard material. By this criterion all of the products are satisfactory.

Experiments designed to evaluate the forces to which weight-bearing casts are subjected during gait need to be carried out before any definitive choice of an alternative splinting material to plaster of Paris can be made. There is little published information about the loading in such a situation, the only really satisfactory study being that of Schenk *et al.* (1969). However, more work is required, as our tests have shown Gypsona to be a relatively weak material and perhaps not best suited to the more demanding splintage requirements.

CONCLUSIONS

Given that an ideal casting material should have a high strength-to-weight ratio, a low flexural modulus (not so low as to produce clinically undesirable deflections), and good radiolucency, then several of the 'second generation' splinting materials satisfy these critera, with the fibreglass products, Scotchcast and Scotchflex, being of note. It may be worth adding a further test to look at the properties of the materials under high strain rate loadings, all of the tests reported here being at strain rates which are low when compared with those to which a lower limb cast is subjected during heel-strike in gait. This may have particular relevance when regarding the suitability of a product for this application.

It is felt that once all the laboratory data are collected, a complete picture of the suitability of the materials can only be made if a clinical evaluation is also included. This will be the next stage in the evaluation of these products.

REFERENCES

British Standard BS2782 (1978). Part 3. Method 335 a. Determination of flexural properties of rigid plastics. British Standards Institute, London.

Gill, J. P. and Bowker, P. (1982). A comparative study of the properties of bandage-form splinting materials. *Engng Med.* **11**, 125–34.

Monro, J. K. (1935). The history of plaster-of-Paris in the treatment of fractures. *Br. J. Surg.* **23**, 257–66.

Schenk, T., Somerset, J. H., and Porter, R. E. (1969). Stress in orthopaedic walking casts. *J. Biomechan.* **2**, 227–39.

Section 2:
Shape Measurement

Regardless of the method chosen, external measurements have three major applications (Moreland *et al.* 1983):

1. To screen populations for the detection of shape abnormality (Aaro and Dahlborn 1981; Willner 1979).
2. To follow the progression of identified deformities (Emans *et al.* 1981; Emans and Hall 1983; Roger *et al.* 1978*a,b*).
3. To assess the effects of various treatment programmes on the progression or correction of identified deformities.

The diagnosis, follow-up, and treatment response in scoliosis is the most well known orthopaedic example of external shape measurements.

A variety of techniques have been used in the detection of scoliosis. Probably the currently most popular method is one or another variation of the moiré fringe technique (Fig. 10.2). Common to all techniques are important issues which relate to the sensitivity and specificity of the measurement. Adair *et al.* (1978), Willner (1979), and others have determined the moiré technique to be sensitive in detecting 85–90 per cent of patients who have significant (greater than 20°) structural scoliosis deformities. In our analysis of ability of moiré fringe to predict the location of a curve, false-positive rate was 5.5 per cent, while the false-negative rate was 11 per cent, giving a predictive value of 88 per cent (Moreland *et al.* 1983). There appears to be general agreement that these screening methods are useful in detecting scoliosis.

The measurement of the progression of deformity (Emans *et al.* 1981; Emans and Hall 1983; Moreland *et al.* 1983; Moreland 1980; Roger *et al.* 1978*a,b*) introduces a number of challenging problems. On one hand, it is desirable to place the patient in some three-dimensional coordinate system. However, too rigid a coordinate system may actually suppress or increase postural changes, while in flexible systems it may be difficult to relate previous to current measurements (Fig. 10.3). The use of computer-simulated 'repositioning' based on standard bony landmarks is one method which has been developed to overcome this problem. The other general problem is to detect progression and clinically important shape change from normal variations which occur as a function of time of day, shoe wear, and fatigue (Roger *et al.* 1978*b*). However, the greatest challenge is to relate internal and external measurements. Emans and Hall (1983) compared progression of idiopathic scoliosis in 140 children as measured by moiré interferommetry, and standard radiographic analyses of Cobb angle. The moiré technqiue was false-negative in 15 per cent, and false-positive in 19 per cent. Moreland *et al.* (1983) found agreement between moiré fringe and Cobb angle in thoracic and thoracolumbar curves was fairly high (regression coefficient = 0.71). The relationship, however, was less predictable for primary lumbar curves (r = 0.42). However, it must be recognized that standard radiographic measurements of lateral deviation (Cobb angle) and rotation (Nash–Moe) have an inherent significant error (Benson *et al.* 1976). It is not clear that

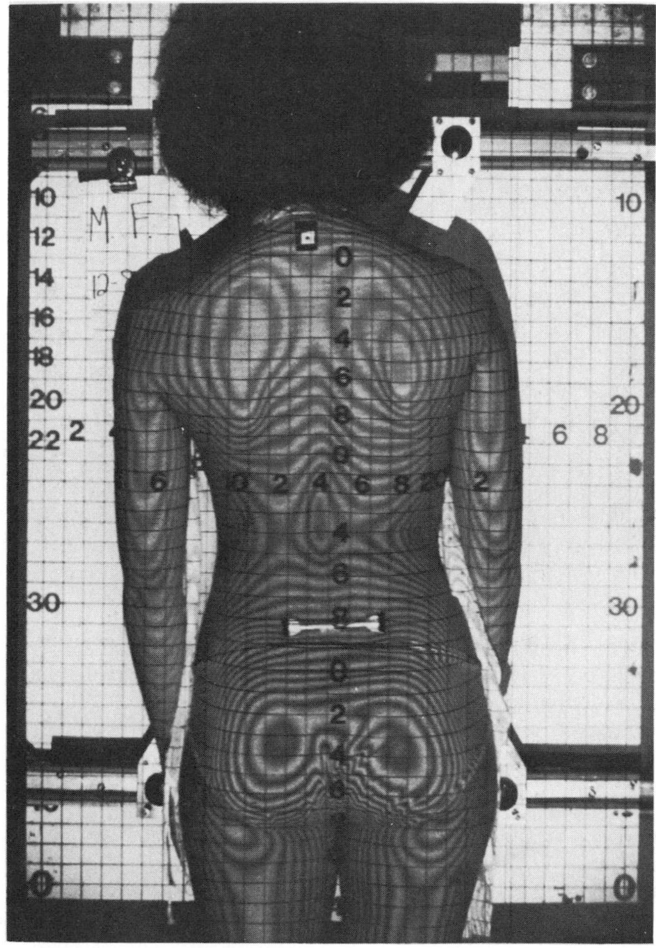

Fig. 10.2. This is a standard moiré fringe topogram made by the interference technique. The contour lines in this picture correspond to 5 mm of elevation.

these radiographic measures will always remain as the 'standard' for description of a scoliosis deformity.

The response of curvatures to treatment programmes has not been assessed fully by topographical measurements. There are many advantages of topographical rather than radiographical analyses of treatment. The tests can be repeated multiple times with no adverse consequences to the patient. They give a measure of cosmetic improvement, and can be used in 'real time', such as in the fitting of electric spinal orthoses.

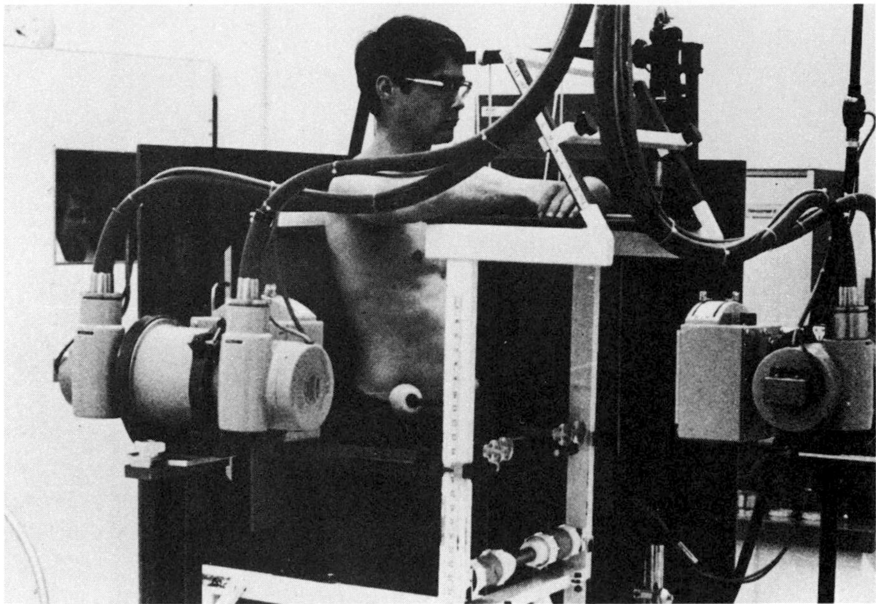

Fig. 10.3. This demonstrates the use of a holding frame for the reproducible measurements of patients with scoliosis. Attempts have been made to have bony landmarks and contacts with the frame without significantly interfering with the positioning of the patient as repeated determinations are made over time, often with a growing child.

RELATIONSHIPS OF EXTERNAL SHAPE TO INTERNAL SHAPE

There are many examples where external shape analysis is important to the assessment of internal shapes. The most dramatic example is the planning of radial keratotomies for correction of myopia (Fig. 10.4). However, the scoliosis deformity again is the most challenging example to be found in musculoskeletal disease. First, an accurate three-dimensional measurement of spinal column shape has yet to be developed. Although CT scan theoretically is most applicable, the radiation hazards have made this method unusable for most research analyses (Adair *et al.* 1978). Stereo radiography has been used effectively as a substitute (Stokes *et al.* 1981). Multiple bony landmarks are identified, and these are used to define position and orientation of each vertebra. These in turn can be related to planes of interest as identified in the external shape measurement. In Chapter 15 Stokes and colleagues demonstrate more accurate relationship between moiré and X-rays may be possible, although significant discrepancies exist, particularly in the measurements of axial rotation by radiographic means (Fig. 10.5). The critical question is which measurement has greater clinical

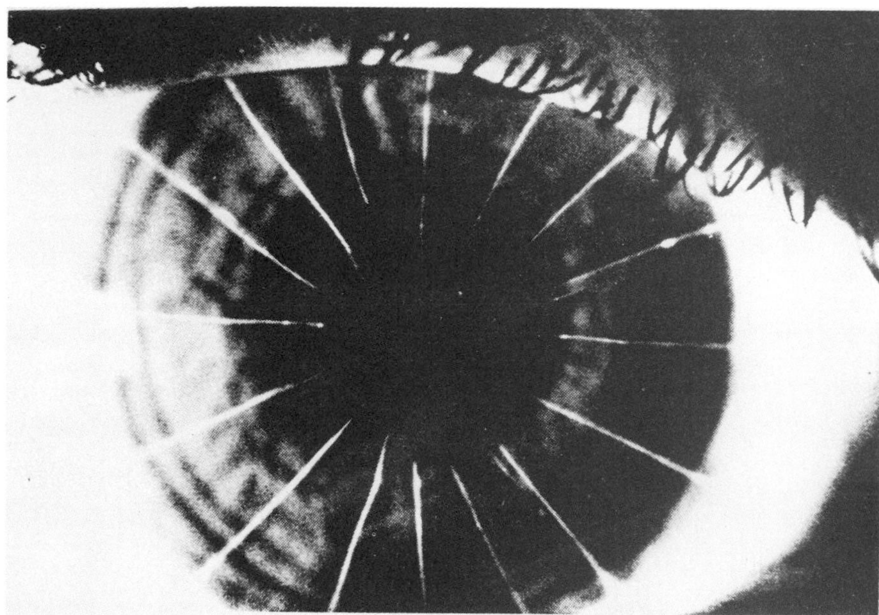

Fig. 10.4. Radial keratotomy performed for myopia. (Reprinted from *Journal of Opthalmic Photography*, Vol. 4, No. 2, Dec. 1981.)

significance (Willner 1981). There is no doubt that curvatures which are at risk for rapid progression such as congenital scoliosis are best evaluated by radiography. In contrast, the major problem of idiopathic scoliosis is cosmetic, except in severe cases where cardiopulmonary function is compromised. Therefore, the external shape measurements and their progression and change may be a more meaningful clinical measurement, particularly in the assessment of the rib hump deformity. The most intriguing possibility is that external shape changes may be a more important predictor of unfavourable cosmetic curve progression than the traditional measures of Cobb angle and Nash–Moe index of rotation.

RELATIONSHIP BETWEEN SURFACE SHAPE AND EXTERNAL ENVIRONMENT

Shape *per se* is not the critical factor to be measured, rather it is the mechanical forces which exist at the interface. Clinicians need no reminder of these forces and how they influence the use of plaster casts, shoe inserts, braces, and seating structures, particularly if there is anaesthetic skin. Computer-controlled machining can now produce precise replications of external shape. Paradoxically, precise replications may produce unfavourable skin pressures, and the reduction of blood flow and tissue oxygen tension (Drummond *et al.* 1982; Reswick

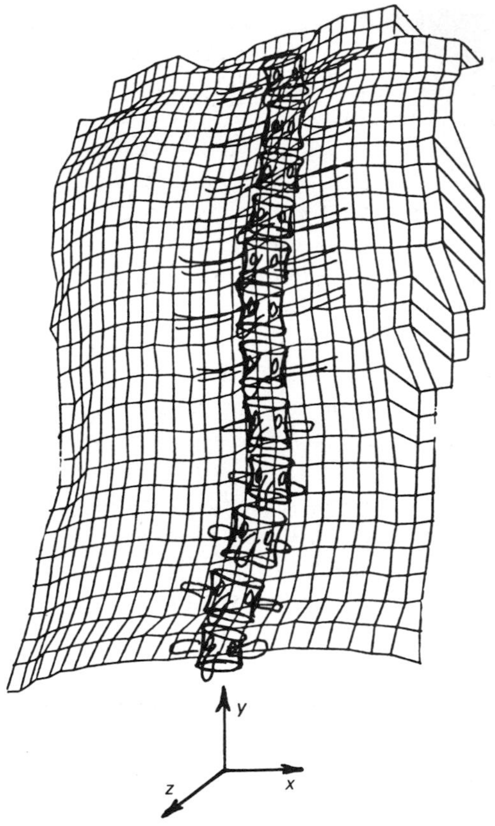

Fig. 10.5. This shows a computer reconstruction of the position of individual vertebral motion segments in a scoliosis deformity, as determined from biplanar radiography. Superimposed are the superficial measurements of surface topography. Even such sophisticated reconstructions show relatively poor correlation between the surface topography, and the internal structural deformity.

and Rogers 1976). Nevertheless, these measurements permit more accurate prosthetic applications, as well as orthoses, cast braces, and the optimization of seating and bedding designs for sensory-impaired patients.

The greatest challenge is to control the internal shape through external shape modifiers. Two examples demonstrate the problem which exists. First, the application of braces is a commonly used method for treating low back disease. The principle use of such braces is to reduce spinal motion, although a secondary beneficial effect may be to increase intra-abdominal pressure. This objective is poorly met by most brace designs. Fidler and Plasmans (1982) have shown significant reductions of flexion/extension spinal mobility do not occur with most braces, with the exception of a single hip spica. Reduction in external spine mobility can be demonstrated by vector stereography but it seems unlikely to translate to significant reductions in the movement of internal vertebral structures. The data of Lumsden and Morris (1968) suggest these apparent reductions in external spinal mobility may in fact be accompanied by increased mobility of spinal motion segments under a number of conditions such as normal walking. In contrast, we do better in the application of braces which seek to reduce or at least allow no progression of existing deformities in growing children. Clinical studies support the efficacy of the Milwaukee and Boston braces in maintaining and often correcting scoliotic deformities; yet there is little information as to how these forces are transmitted to the vertebral structures. There are some research data which suggest that under certain conditions the external forces may in fact translate to unfavourable internal forces. For example, Moreland (1980) has shown that derotation devices which apply torsion to the lower legs of experimental animals primarily affect ligamentous structures. Only very high force application produced structural change in bone, which occurred principally at the epiphyseal plate. In our laboratories, we are attempting to understand this relationship between internal structure, external shapes, and forces by analysing the effects of various orthoses on the foot arch development of the child (Figs. 10.6–10.8). This type of study brings together internal measure of shape, external measures of shape by moiré fringe topography, and force analyses as they occur at the orthotic-skin interface.

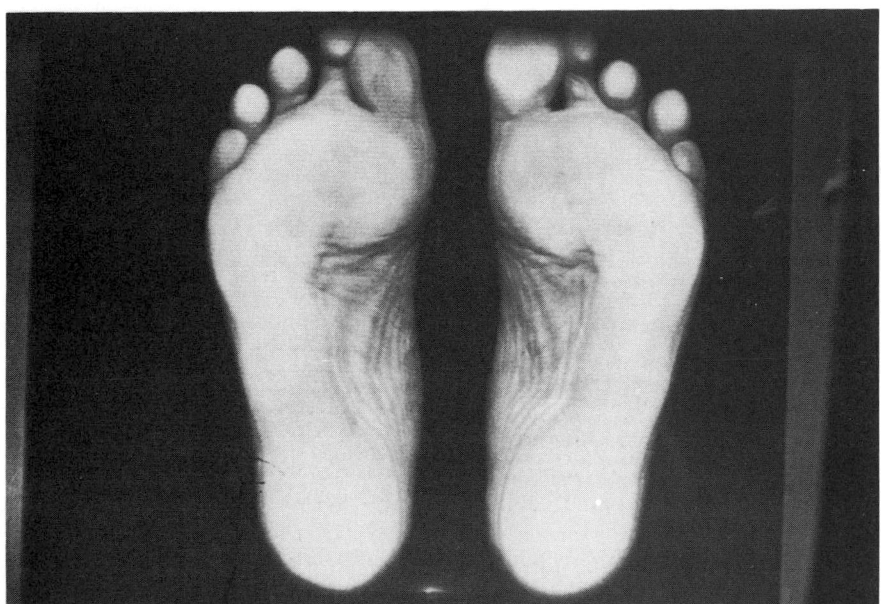

Fig. 10.6. This moiré topogram of a plantar aspect of the foot demonstrates the type of shape measurement possible. Accurate replications can occur with the standing posture and the arch of the foot development followed.

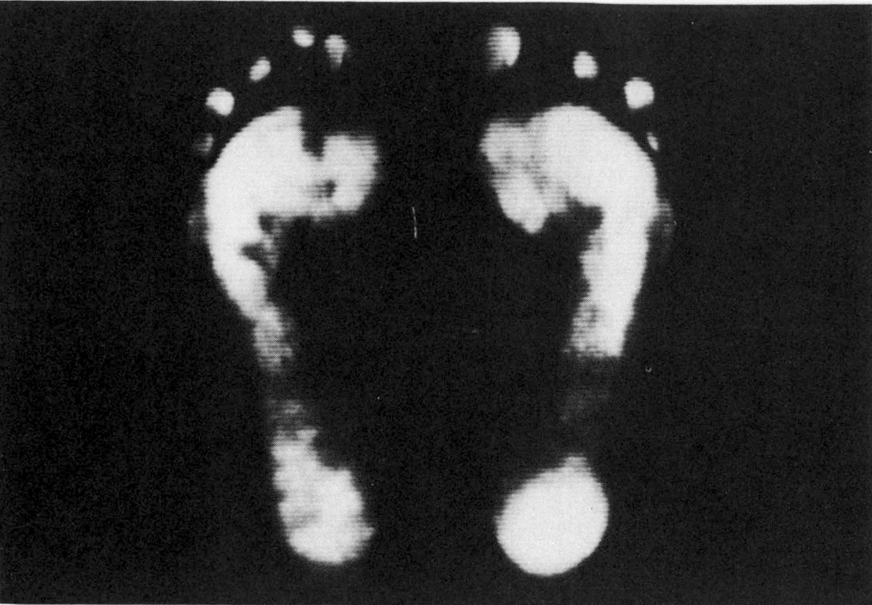

Fig. 10.7. A pressure device using pressure sensitive film which in combination with the moiré fringe topogram gives both mechanical and topographical data, and permits in other conditions (for example, prominent metatarsal heads) the determination of how shape affects pressure and how appropriate modifications can be made in shoe design for treatment.

SUMMARY

It is clear we have progressed in 'shaping up' our understanding of the measurement of shape. The shape of things to come is exciting, and presents many challenges as we seek more precisely to measure and control the internal and external shapes of multiple structures in the human body.

REFERENCES

Aaro, S. and Dahlborn, M. (1981). Estimation of vertebral rotation and spinal and rib cage deformity in scoliosis by computer topography. *Spine* **6**, 460–7.

Adair, I. V., Van Wijk, M. C., and Armstrong, G. W. D. (1978). Moiré topography in scoliosis screening. *Clin. Orthopaed.* **129**, 165–71.

Bender, T. (1976). *Environmental design primer*. Schocken Books, New York.

Benson, D. R., Schultz, A. B., and DeWald, R. L. (1976). Roentgenographic evaluation of vertebral rotation. *J. Bone Jt Surg.* **58A**, 1125–9.

Burwell, R. C. (1978). Biostereometrics shape replication and orthopaedics. In *Orthopaedic engineering* (ed. J. D. Harris and K. Copeland) pp. 51–85. The Biological Engineering Society, London.

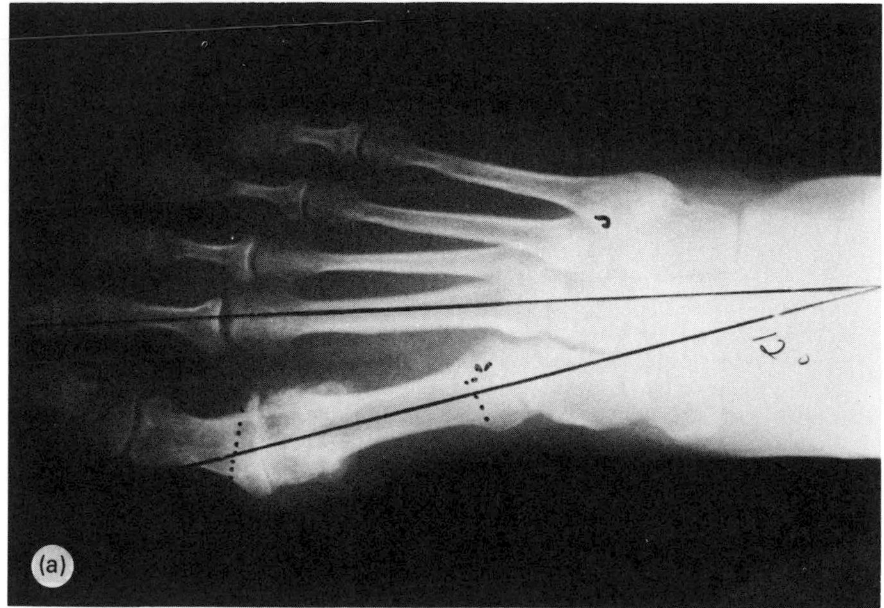

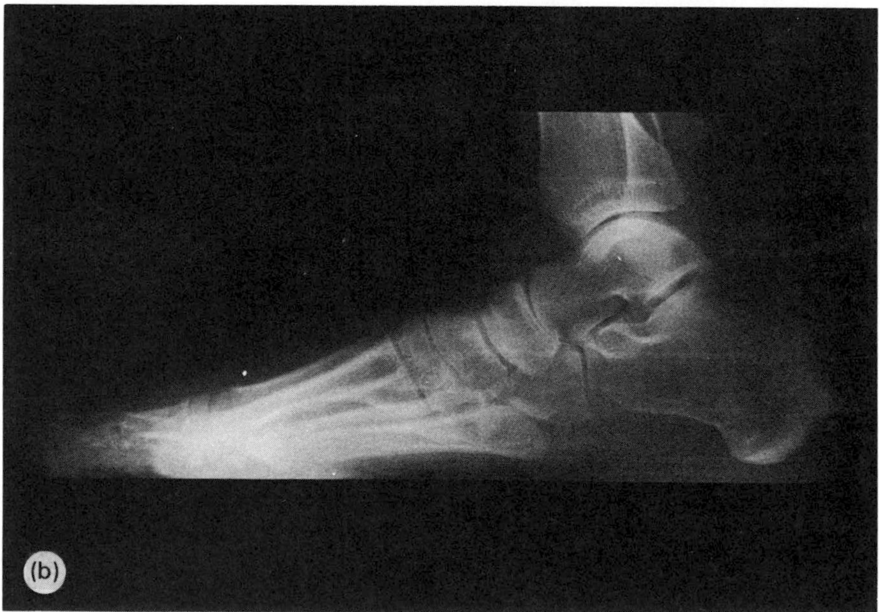

Fig. 10.8. AP and lateral radiographs of the foot, demonstrating the various radiographic measurements which are made and correlated with the moiré fringe topogram.

Drummond, D. S., Narenthania, R. G., Rosenthal, A. N., Breed, A. L., Lange, T. A., and Drummond, K. D. (1982). A study of pressure distribution measured during balanced and unbalanced sitting. *J. Bone Jt Surg.* **64A**, 1034–9.

Emans, J., Bailey, T., and Hall, J. (1981). Preliminary observations in the longitudinal followup of mild idiopathic scoliosis utilizing shadow Moiré topography. In *Moiré fringe topography and spinal deformity* (ed. M. S. Moreland, M. H. Pope, and G. W. D. Armstrong) pp. 166–70. Pergamon Press, New York.

— and Hall, J. E. (1983). Detection of progression in idiopathic scoliosis by Moiré shadow topography. Paper #145, 50th Annual Meeting. American Academy of Orthopaedic Surgeons.

Fidler, M. T. and Plasmans, C. (1982). Effects of external supports on the mobility of the lumbosacral spine. Presented at the 8th Annual Meeting of the International Society for Study of the Lumbar Spine. Toronto.

Lumsden, R. M. and Morris, J. M. (1968). An *in vivo* study of axial rotation and immobilization at the lumbosacral joint. *J. Bone Jt Surg.* **50A**, 1591–602.

Moreland, M. S. (1980). Morphological effects of torsion applied to growing bone. *J. Bone Jt Surg.* **62B**, 230–7.

—, Pope, M. H., Stokes, I. A. F., and Weierman, R. (1983). Diagnostic uses of Moiré fringe topograms. Paper #144, 50th Annual Meeting, American Academy of Orthopaedic Surgeons.

Reswick, J. B. and Rogers, J. E. (1976). Experience at Rancho Los Amigos Hospital with devices and techniques to prevent pressure sores. In *Bedsore biomechanics* (ed. R. M. Kennedy, J. M. Cowden, and J. T. Scales) pp. 301–10. Macmillan, London.

Roger, R., Frymoyer, J. W., and Stokes, I. A. F. (1978*a*). Investigation of changes in the posture of the spine. The effects of heel raise and the working day. In *Orthopaedic engineering* (ed. J. D. Harris and K. Copeland) pp. 86–91. The Biological Engineering Society, London.

—, Stokes, I. A. F., Harris, J. D., Frymoyer, J. W., and Ruiz, C. (1978*b*). Monitoring adolescent idiopathic scoliosis with Moiré fringe photography. *Engng Med.* **8**, 119.

—, —, —, and Turner-Smith, A. R. (1981). Monitoring scoliosis with serial measurements of three-dimensional back shape. In *Moiré fringe topography and spinal deformity* (ed. M. S. Moreland, M. H. Pope, and G. W. D. Armstrong) pp. 258–66. Pergamon Press, New York.

Stokes, I. A. F., Wilder, D. G., Frymoyer, J. W., and Pope, M. H. (1981). Assessment of patients with low-back pain by biplanar radiographic measurements of intervertebral motion. *Spine* **6**, 233–40.

Sullivan, L. H. (1976). The testament of stone. In *Environmental design primer* (ed. T. Bender) p. 117. Schocken Books, New York.

Willner, S. (1979). Moiré fringe topography for the diagnosis and documentation of scoliosis. *Acta orthopaed. scand.* **50**, 295–302.

— (1981). Comparison between Moiré and X-ray findings in structural scoliosis. In *Moiré fringe topography and spinal deformity* (ed. M. S. Moreland, M. H. Pope, and G. W. D. Armstrong) pp. 157–65. Pergamon Press, New York.

11 Imaging the Form of the Back with Airborne Ultrasound

L. MAURITZSON, G. BENONI,
K. LINDSTRÖM, AND S. WILLNER

INTRODUCTION

Many different non-invasive methods can be used to achieve a three-dimensional description of the shape of a body, e.g. moiré topography. Mechanical equipment such as spinal pantographs and kyphometers have been developed, but using these it is necessary to touch the body. In order to develop a non-contact method we have therefore started experimenting with airborne ultrasound. Distance measuring with airborne ultrasound is utilized in nature by several animals who navigate and locate food by it with great success.

METHOD

A short ultrasound pulse with a frequency of 50 kHz is emitted from an electrostatic ultrasonic transducer which is directed towards the patient's back. The sound wave progresses in a beam to the back with a velocity, c m/s, that is dependent on the ambient temperature:

$$c = 331.6 + 0.6\,T,$$

where T is temperature in °C (Lindström *et al.* 1982).

The echo reflected from the surface of the back is received with the same transducer, which is now used as a microphone. The measured propagation time between start of the emitted pulse and the received and detected echo is proportional to the distance between the transducer and that particular point where the echo was reflected.

The relation between the distance x and the measured propagation time t is

$$t = \frac{2x}{c}.$$

Although the sound velocity is dependent on the ambient temperature, the axial resolution for the distance measurement is between 0.1 mm and a few millimetres, provided that the system is calibrated during the measurement either with a built-in testblock or by measuring the temperature.

First we placed a single ultrasonic pulse-echo system directed towards the patient's back and made a vertical scan with the system. This resulted in a continuous scanning of the back, along the spine for example. The scanning time was about 20 seconds. We got a lot of artefacts from movement of the patient, and the reconstructed picture displayed on an oscilloscope was not easy to quantify. Therefore we have developed a multi-element linear array with 16 single transducers, each with a diameter of 40 mm, stacked on each other with 50 mm between each. The distance between the transducer and the body is in the range of 80 mm and the divergency of the soundbeam is negligible. A control circuit is constructed to share the time between the 16 transducers and to use them as single distance-measuring system. This information is placed in a semiconductor memory and at the same time displayed on a monitor, for the operator, as a 16-point profile. As the time for making a total scan only takes 30 ms, we do not get any artefacts from the movements of the patient, but instead we get a real-time imaging of the movements. When analysing the image, the information is sent to a personal computer (HP–85) where it is presented as a table of distance values, or can be illustrated by a profile, where some points are used as reference. The computer can also revise the information, for example calculate the difference between the two profiles and present the result.

APPLICATIONS

The cases presented here are from a pilot study. During this study, we have made several modifications to the apparatus as well as to the procedures. We therefore prefer to illustrate this method by discussing two cases.

The patients stand with their back towards the array of transducers, as shown in Fig. 11.1. In order to standardize the posture of the patients we use a positioning frame with two horizontal nylon strings which are not detected by the scanner.

The patients stand relaxed and lightly touching the nylon strings with the most prominent part of the scapulae and the buttocks. Through this arrangement the position of the patient is similar to that used when taking moiré photographs (Willner and Willner 1982). We can scan the back of the patients when they are standing in a relaxed position and only touching the nylon strings with the buttocks in order to detect any inherent rotation of the trunk.

A horizontal ruler across the positioning frame is adjusted so that its lower edge is level with the spinal process of vertebra C7. Aided by a vertical laser plane producing a visible line on the back, the array of transducers is moved to the desired position. So far we have scanned most patients along five vertical lines:

– one midline (M), through the spinal process of vertebra C7
– one line 5 cm to the left (L5) of the midline
– one line 5 cm to the right (R5) of the midline

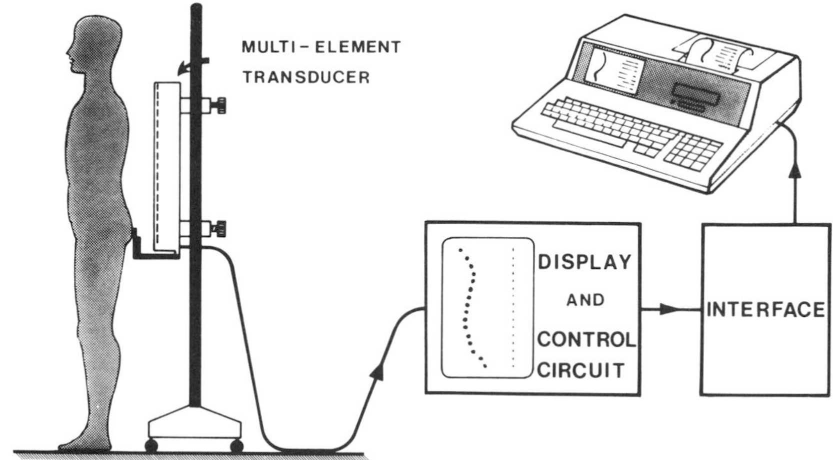

Fig. 11.1. Imaging the back with a linear array ultrasound scanner for use in air.

– one line through the inferior angle of the left scapula (LS)
– one line through the inferior angle of the right (RS) scapula.

Each ultrasound beam has a diameter of 40 mm and the scanner registers the first echo it detects after each sound pulse. It thus records the nearest point on the object within this 40 mm circle. The recorded point is not necessarily situated in the centre of the circle but on the display it is shown as if the echo comes from a point exactly opposite to the midpoint of each transducer.

The first case has a practically symmetrical moiré pattern and the scans on both sides of the back are symmetrical (Fig. 11.2).

In the second case, with scoliotic asymmetry of the back, these scans are asymmetric. The computer can be programed to use the available data in several ways, for example evaluating the kyphosis and lordosis angles. One simple way of expressing the asymmetry of the back, as seen in structural scoliosis, is to subtract corresponding scans on both sides from each other and to plot the difference in a diagram (Fig. 11.3).

In cases with symmetrical backs, these points would lie on a straight line as demonstrated in the first case, to the left in Fig. 11.3. In the second case, with the scoliosis, the difference shows segmental changes, to the right in Fig. 11.3.

From the moiré pictures it is possible to construct contours of the back along the scanning lines and to compare these contours with those obtained by the scanner. Although the patient has moved between the scanning and the taking of the moiré photographs, the conformity of the scanning contour to the moiré contour is good.

However, the scanner has a tendency to straighten the curves owing to the fact that the ultrasound beam has a finite width. This can be minimized by using

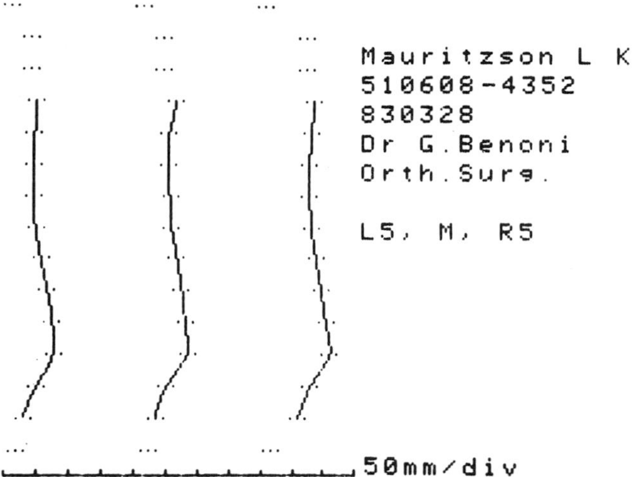

Fig. 11.2. Computer presentation of a practically symmetric back. L5, M, and R5 denote the scan (see text). Each triplet of points represents one transducer. The two lowermost and the three uppermost transducers are reference points. For technical reasons the lowermost transducer on the array is not displayed.

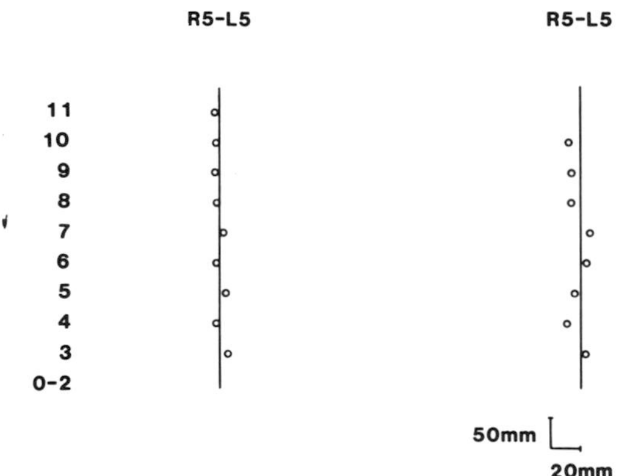

Fig. 11.3. The L5 scan subtracted from the R5 scan in a patient with a practically symmetric back, left, and in a patient with scoliosis, right. The numbers to the left represent the transducers, numbered from below. Reference transducers are omitted.

a correction factor when doing the calculations. An improvement would be to use smaller transducers with a higher frequency, to produce ultrasound beams with decreased diameters. This requires more sensitive amplifiers with a higher signal-to-noise ratio. Another improvement would be a scanner with three or more linear arrays that could scan the whole back in one single measurement.

CONCLUSION

We believe that this method of making measurements is worth further development. For example, it can be used to measure the size and volumes of objects. So far we have not exploited the real-time ability of the scanner to record movements, but our view is that this method can be applied to many fields of biomechanical measurement.

REFERENCES

Lindström, K., Mauritzson, L., Benoni, G., Svedman, P., and Willner, S. (1982). Application of air-borne ultrasound to biomedical measurements. *Med. biol. Engng Comput.* **20**, 393–400.

Willner, S. and Willner, E. (1982). The role of Moiré photography in evaluating minor scoliotic curves. *Int. Orthopaed.* **6**, 55–60.

12 Shape Measurement in the Scoliosis Clinic

A. R. TURNER-SMITH AND DEREK HARRIS

INTRODUCTION

Scoliosis is an unpleasant deformity which is severe enough to warrant assessment and possible treatment of about 0.3 per cent of the population (Keim 1979; Desmet *et al.* 1981). The aetiology of about 70 per cent of cases, referred to as 'idiopathic', is unknown. In these cases, progression of the disease is usually most rapid during the adolescent growth spurt, during which period it is not unusual to take several radiographic series each year. The majority of patients followed in scoliosis clinics, however, never reach the point where surgical intervention is necessary. Even when surgery is performed, it is often aimed at improving cosmetic appearance as much as improving the mechanical stability of the spine and function of the pulmonary system. Indeed, there is an opinion that minimal surgical intervention is best for all but severe cases, and that treatment is desirable chiefly for cosmetic appearance (Weinstein *et al.* 1981). Thus, in addition to the assessment of the spinal deformity by radiography, a thorough programme for monitoring scoliosis will include the measurement of cosmetic appearance. Not only does the back shape give an indication of the severity of the disease, but a change of shape will reflect the effectiveness of treatment (exercise, bracing, etc.) and underlying skeletal changes.

THE PRESENTATION OF SHAPE DATA

The eye is a sensitive gauge of shape, but it is not possible to retain an exact enough memory of the appearance of a scoliotic back to recognize a small but significant deterioration in deformity. Various quantitative techniques have been proposed to assist the clinician. Direct tactile methods include measurement with fixed or flexible rules and deformation or movement of probes. These methods require the patient to remain still for a period of time while the measurements are made. In general they are awkward to apply and very dependent on the skill and consistency of the operator.

Non-tactile methods usually use optical means, either photographic or electronic, to image the shape of the back. Moiré topography is the most popular of these and a variety of methods have been devised (Meadows *et al.* 1970; Takasaki 1970, 1981; Herron 1972; Lovesey 1973; Joel 1974; Willner 1979; Hierholzer

Fig. 12.1. A typical moiré topograph of a right thoracic and left lumbar scoliosis.

and Frobin 1981; Suzuki *et al.* 1981; Ishida *et al.* 1982). Moiré topography can be produced quickly, is completely safe, and the appearance, representing a contour 'map' of the back, is subjectively convincing (Fig. 12.1).

Two particular problems are associated with the use of moiré topography. First, the fringe contours of the moiré topography show not only the shape of the back, but the posture of the patient. If this position changes even slightly, the appearance of the picture is markedly changed. For comparative study therefore the patient must be rigidly constrained. In doing this there is the possibility that, in addition to the normal variations of stance which occur from time to time, a significant postural alteration caused by the deformity will be eliminated. Second, the contour picture itself does not provide numbers which can be compared. Various methods have been proposed to assist in the analysis of moiré topographs and to quantify the deformities due to scoliosis (Willner 1979; Suzuki *et al.* 1981; Ishida *et al.* 1982; Moreland *et al.* 1981; Drerup 1972, 1981; Shinoto *et al.* 1981; Kamal 1983), but the measurement and analysis procedures tend to add time and complication to the moiré technique, nullifying its main attractions of speed and simplicity.

Opto-electronic methods of shape measurement are being developed in many centres (television systems, live-scan cameras, etc.) which provide shape data direct to a computer for analysis. The detailed method of recording shape, however, is secondary to the fundamental aim of analysing and presenting shape information for clinical use. Such a presentation will contain two types of information: (i) numerical information quantifying parameters relevant to the patient's disease; and (ii) graphical information which pictorially describes and records the appearance of the scoliosis. Both types of information should be easily compared from one time to the next. The ideal numeric and graphical description of scoliosis shape will be sensitive to shape change due to deterioration or improvement in the scoliosis, but independent of transient postural changes. It will be independent of the method of shape measurement (moiré topography, electronic imaging, etc.) and it will be expressed in terms which are significant to the clinician rather than as some obscure mathematical function.

The presentation shown in Fig. 12.2 is an attempt to meet these requirements. It has been used regularly in the scoliosis clinic of the Nuffield Orthopaedic Centre over the last two years when it has been applied to shape data collected by both a moiré photographic method followed by manual digitizing, and by a direct television-to-computer method, 'ISIS', outlined at the end of this chapter (Turner-Smith *et al.* 1981; Turner-Smith 1983).

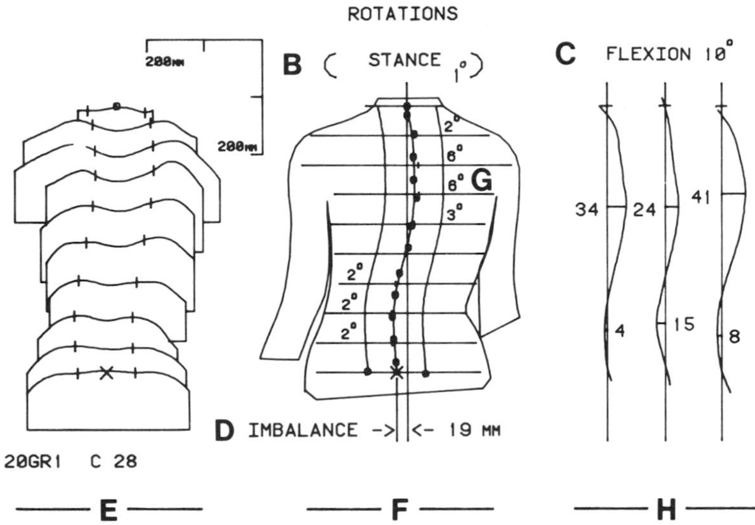

Fig. 12.2. Shape analysis of a right thoracic idiopathic scoliosis.

In the clinic, patients are marked with black adhesive stickers over the vertebra prominens (C7/T1) and dimples of Venus (posterior superior iliac spines, PSIS). Coordinates of the back surface are adjusted to be relative to these landmarks:

1. The coordinate origin is set at the vertebra prominens.
2. A point half way between the dimples of Venus is referred to as the 'sacrum point'. The angle of the back surface in the transverse plane at this sacrum point defines the orientation of the coronal plane (Fig. 12.3). This corrects for differences in the angle at which the patient stands with respect to the observer.

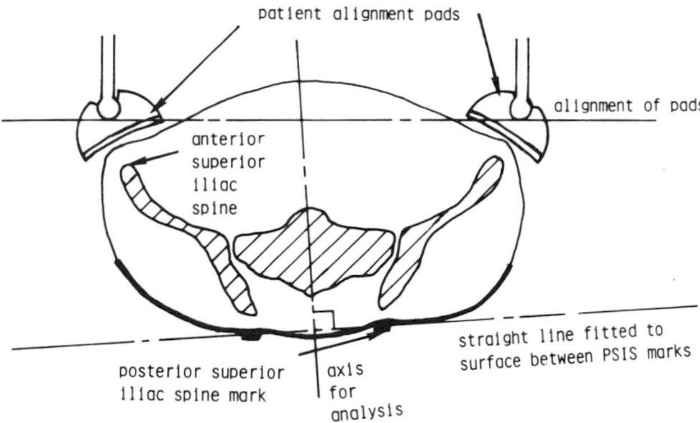

Fig. 12.3. Location of patient, transverse plane.

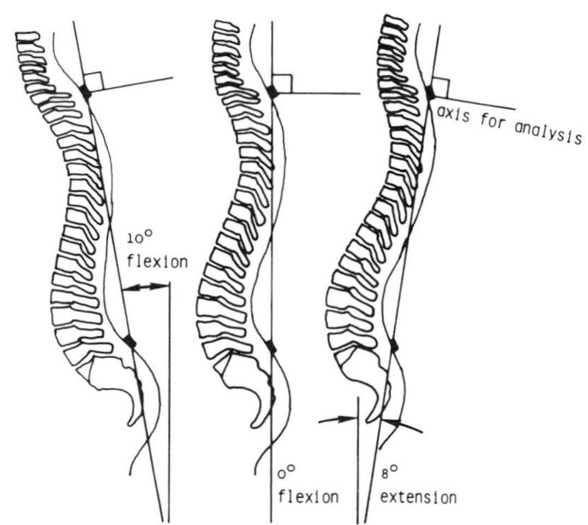

Fig. 12.4. Location of patient, sagittal plane. Skin markers are over vertebra prominens and PSIS.

3. The sacrum point is set directly beneath the vertebra prominens in the sagittal plane (Fig. 12.4). This provides a simple correction for differences in flexion or extension of the spine.

The line of the palpable spine is also marked. From this line, profiles are drawn and an area is defined whose shape is expected to have a correlation with the orientation of the underlying spine, ribs, and associated musculature.

Marking the patient in this way takes only a few minutes and is judged a more reliable method of identifying repeatable landmarks than looking for surface features by shape alone.

FEATURES OF THE PRESENTATION (FIG. 12.2)

The heading (A) identifies the month and year in which the picture was taken, together with the patient code and hospital number. 'Stance' (B) is the correction which has had to be applied to compensate for the angle at which the patient stands with respect to the observer, as described above. An angle appearing on the right indicates an apparent rotation of the subject turning to the right, with respect to the anterior support pads. An angle appearing on the left indicates a rotation to the left. Flexion (C) is the angle to the vertical of a line from the vertebra prominens to the sacrum point. Imbalance (D) is the lateral distance in millimetres of the sacrum point to the vertical dropped from the vertebra prominens. Sections (E): ten horizontal sections through the back are shown at equal distances between the vertebra prominens and the sacrum point. Growth of the subject between examinations is thus automatically compensated. The two vertical marks on each section are the same distance apart as the PSIS, but centred on the apparent line of the spine at the level of the section. The average slope of the section between these two marks is the measure of rotation discussed under (G) below. Body outline (F): the 10 horizontal sections are marked on the body outline and appear as straight lines in this view. The PSIS, the line of the spine, and the vertebra prominens are marked by dots. The sacrum point is marked with a cross. The line of the spine is also followed up from the PSIS marks on either side of the spine. The two outer lines define a surface whose shape over its entire length, from sacrum to neck, is determined chiefly by the underlying spine, ribs, and associated musculature. Rotations (G): the angle marked on each horizontal section on the body outline is the rotation of the surface represented by the slope of the section shown on the left between the vertical marks. An angle representing a hump on the right is printed on the right; an angle representing a hump on the left is printed on the left. Vertical profiles (H) are measured along three lines following the shape of the spine shown on the body outline. This is important since if, for example, a profile were measured along a straight line from vertebra prominens to sacrum, at heights where the spine deviated from perfect straightness, the profile would show muscle bulk rather than true shape over the spinous processes. By measuring along the line of the spine this distortion is avoided. The maxima and minima of these curves represent the maxima of the kyphosis and lordosis, and their magnitude and position are marked with figures and horizontal lines.

Repeatability of the analysis system has been tested by measuring a replica of a scoliotic back presented at various angles (as it were in different postures). For a 20° change in 'stance' the shape of the sections and profiles changes hardly at all, the lordosis and kyphosis sizes change by 1 mm and the surface angles about the spine change by 2°. The repeatability of the analysis system is thus greater than the repeatability of patient posture.

Between one clinic and the next, back shape changes, due to growth or treatment, may make it difficult to compare the appearance of the subject,

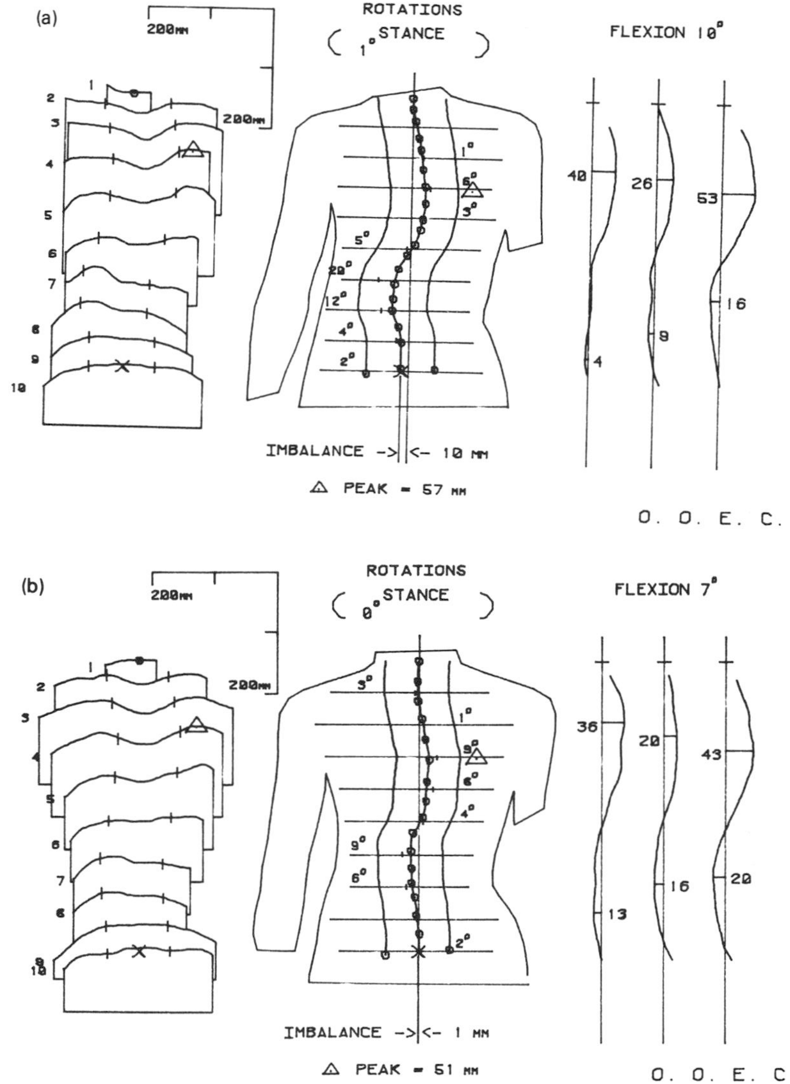

Fig. 12.5. (a) Lumbar scoliosis before bracing, Cobb angle 56°. (b) Lumbar scoliosis after bracing, Cobb angle 62°.

but, the changes are readily seen in the shape analysis, even when the radiograph shows little or no change, as was the case in Fig. 12.5. This highlights the importance of shape measurement which can provide objective evidence for the effectiveness of treatments which otherwise have little impact on the radiographic appearance of the back. A recent short series of 22 measurements on eight patients with thoracic scoliosis indicated that as simple a measure as the surface angle at the level of maximum deformity was a good indicator of improvement or deterioration in clinical condition (Fig. 12.6).

The costaplasty operation for severe scoliotic deformity is designed to improve appearance of the back. Comparison of back shape analyses before and after operation (Fig. 12.7) shows the precise cosmetic effect of the operation in a way which no other technique in use provides.

SURFACE ANGLE ASSESSMENT	Curve	CLINICAL ASSESSMENT	
		Controlled	Progressing
	Controlled	11	2
	Progressing	3	6

Fig. 12.6. Idiopathic thoracic scoliosis, 22 measurements from eight patients, six months or more between measurements.

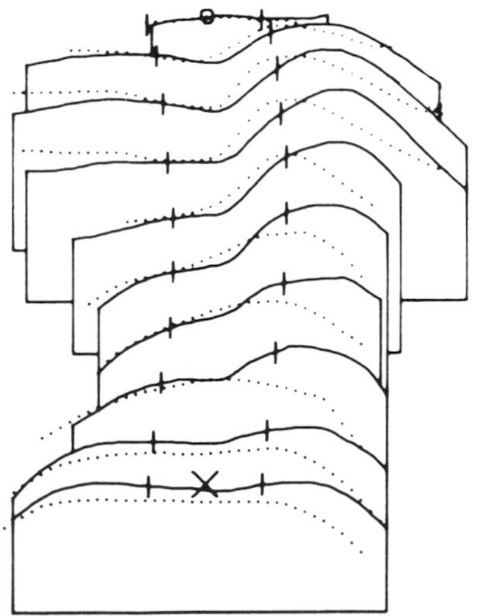

Fig. 12.7. Shape analysis pre and post costaplasty. The plain section lines are calculated from a moiré topograph before operation. The dotted lines are from the post-operative study.

A FAST METHOD FOR BACK SHAPE ANALYSIS IN THE CLINIC—ISIS

The method of moiré topography and manual digitizing used at the Oxford Orthopaedic Engineering Centre (Turner-Smith *et al.* 1981) has been useful for research but has failed as a clinical tool, first because the effort required for film processing, manual digitizing and computer processing is more than can be supported (about 30 minutes per patient), and secondly because the measurements are required by the physician at the clinic while the patient is present, rather than some days later.

A method of direct computer input of shape has been developed by which a patient can be measured and an analysis of their scoliosis produced within seven minutes (Turner-Smith 1983). The instrumentation is shown diagrammatically in Fig. 12.8. A standard 35 mm slide projector and a television camera are mounted rigidly one above the other in a frame which is free to rotate about a horizontal axis. A horizontal line of light from the projector is viewed by the television camera, mounted on its side and at an angle of about 30° to the projector axis. Thus, as it falls on an object, the horizontal fan of light from the projector appears to be distorted as the surface of the object lies closer to or further away from the optical axis of the projector. The scan of a human back takes less than two seconds, during which time television data are stored directly by a Digital Equipment Corporation PDP–11/23 computer. Analysis of the data then proceeds to produce a computer file of back shape information and a printed analysis in the same form as shown in Fig. 12.2.

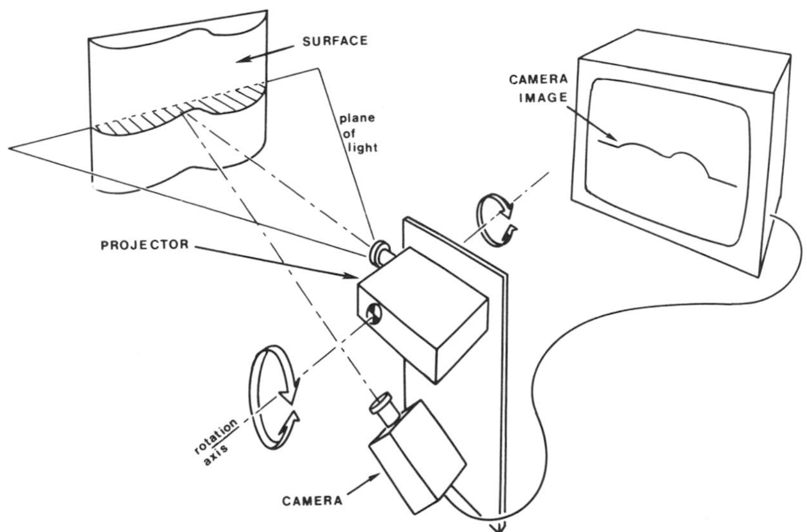

Fig. 12.8. Principle of operation of the 'ISIS' shape measurement equipment.

Shape data produced in this way may be compared directly with data from digitized moiré pictures, and once stored in digital form, any method of shape analysis is possible.

CONCLUSION

The importance of shape measurement in the assessment of scoliosis has been emphasized. How the shape of the scoliotic back is best described is still uncertain. One method of presentation has been presented in detail; it is easy to understand and provides direct numerical comparison between data from one clinic to the next. Two methods of shape measurement have been described: a photographic moiré technique and a television technique, ISIS. Both methods yield the same data, but the latter can provide the clinician with shape analysis within a few minutes.

It is to be hoped that the ease with which shape can now be measured and analysed will enable this important information to play a valuable part in the understanding and treatment of scoliotic deformity.

REFERENCES

Desmet, A. A., Cook, L. T., and Tarlton M. A. (1981). Assessment of scoliosis using three-dimensional radiographic measurements. *Automedica* **4**, 25-36.

Drerup, B. (1972). Contourometric evaluation of moiré topograms by nomographic techniques. *Engng Med.* **11**, 33-8.

— (1981). The measurement of angles in moiré topograms. In *Moiré fringe topography and spinal deformity*, pp. 190-200. Pergamon, Oxford.

Herron, R. E. (1972). Biostereometric measurement of body form. *Yearb. phys. Anthropol.* **16**, 80-121.

Hierholzer, E. and Frobin, W. (1981). Rasterstereographic measurement and curvature analysis of the body surface of patients with spinal deformities. In *Moiré fringe topography and spinal deformity*, pp. 267-76. Pergamon, Oxford.

Ishida, A., Suzuki, S., Imai, S., and Mori, Y. (1982). Scoliosis evaluation utilising truncal cross-sections. *Med. biol. Engng Comput.* **20**, 181-6.

Joel, H. C. (1974). The Corpograph—a simple photographic approach to three-dimensional measurements. In *Biostereometrics 74. Proceedings of the Symposium of Commission V.* Washington, DC.

Kamal, S. A. (1983). Determination of degree of correction spinal deformity by moiré topographs. In *Second International Symposium on Moiré Topography and Spinal Deformity*, Munster, pp. 117-24. Fischer, Berlin.

Keim, H. A. (1979). Scoliosis. *Clin. Symp.* **30**, 2-30.

Lovesey, E. J. (1973). A simple photographic technique for recording three-dimensional head shape. *Med. biol. Illust.* **23**, 210-13.

Meadows, D. M., Johnson, W. O., and Allen, J. B. (1970). Generation of surface contours by Moiré patterns. *Appl. Optics* **9**, 942-7.

Moreland, M. S., Barce, C. A., and Pope, M. H. (1981). Moiré topography in scoliosis: pattern recognition and analysis. In *Moiré fringe topography and spinal deformity*, pp. 171-85. Pergamon, Oxford.

Shinoto, A., Ohtsuka, Y., Inoue, S., Idesawa, M., and Yatagai, T. (1981). Quantitative analysis of scoliosis and kypyosis deformity by moiré method. In *Moiré fringe topography and spinal deformity*, pp. 206–24. Pergamon, Oxford.

Suzuki, N., Armstrong, G. W. D., and Armstrong, J. (1981). Application of moiré topography to spinal deformity. In *Moiré fringe topography and spinal deformity*, pp. 225–40. Pergamon, Oxford.

Takasaki, H. (1970). Moiré topography. *Appl. Optics* **9**, 1467.

— (1981). Moiré topography and its application to human body. In *Moiré fringe topography and spinal deformity*, pp. 1–17. Pergamon, Oxford.

Turner-Smith, A. R. (1983). Television scanning technique for topographic body measurements. In *Biostereometrics 82. SPIE Proceedings*, Vol. 361, pp. 279–83.

—, Harris, J. D., Abery, J. M., and Osborne, M. (1981). Back shape measurement. *Oxford Orthopaedic Engineering Centre: Annual Report* No. 8, pp. 41–9.

Weinstein, S. L., Zavala, D. C., and Ponseti, I. V. (1981). Idiopathic scoliosis (long term follow-up and prognosis in untreated patients). *J. Bone Jt Surg.* **63A**, 702–12.

Willner, S. (1979). Moiré topography for the diagnosis and documentation of scoliosis. *Acta orthopaed. scand.* **50**, 295–392.

13 The Application of Optical Contour Mapping to the Recumbent Patient

M. D. L. MORGAN, A. J. PEACOCK, A. R. GOURLAY, AND D. M. DENISON

INTRODUCTION

The study of lung function usually emphasizes the mechanical or gas exchanging properties of the lungs or of gas flow through its airways. Often, little thought is spared to the muscles and skeletal elements that generate the power to breathe. Over the past two years we have been interested in what can be learnt about the function of these elements as well as the lungs themselves by studying the surface of the chest and abdomen during breathing. Observation of the motion of the chest wall is, of course, regular practice during clinical examination and often aids diagnosis of underlying disease. Our aim has been to develop a method of making quantitative and detailed analysis of these movements. The method that we have chosen, optical contour mapping, has been developed at the Brompton Hospital in association with IBM (UK) (Peacock *et al.* 1984).

The application of this system has so far been limited to those subjects or patients who are fit enough to come to the Lung Function Laboratory and stand upright in the correct position. It would be more valuable to study those patients who are unable to come to the Lung Function Laboratory or those who are seriously ill and bed-bound. For that reason we have entered into a collaborative study with the National Spinal Injuries Centre at Stoke Mandeville to develop optical contour mapping equipment capable of studying the recumbent patient and use it to make a special study of those patients with high spinal injury and tetraplegia (Morgan *et al.* 1984).

METHODS

The principle of the method is that two fixed line patterns are projected on to the body from either side. The pattern is distorted to appear as contour lines when viewed by two cameras at 90° to the axis of projection. The position of the patterns with reference to the centre of the system is known exactly. The image obtained is digitized and the apparent three-dimensional coordinates can be corrected for the divergent optics of the projector and convergent optics of the camera to give the real coordinates. The information from both front and back can be used with a choice of program to calculate contained trunk volume or used to generate cross-sectional profiles.

There are special considerations to studying respiration in patients lying down. No view of the back of the patient can be obtained unless a transparent bed is used. So an assumption must be made that all respiratory movement must be forwards or sideways of the firm bed. When two projectors are used to illuminate the patient in a horizontal plane a certain central portion of the patient is in shadow (Fig. 13.1). This 'no-man's area' provides no information about the surface and could be eliminated if the projectors shone down at a slight angle on to the body. Such an arrangement, though, would prevent the use of available programs which use right-angle geometry, and also introduce the problem of distinguishing contours in one field overlapped by contours from the other projector's field. In practice this small area of shadow can be reduced by rotating the patient slightly, but in most cases the area is small enough to be ignored.

The fields of view of two cameras would cover more of the body and provide more information, but there are many advantages to using a single overhead camera to capture the image. The distortion of the pattern and therefore the sensitivity is greatest if viewed at 90° and again, the same programs can be used if the geometry is maintained. Two images require twice as much analysis, but

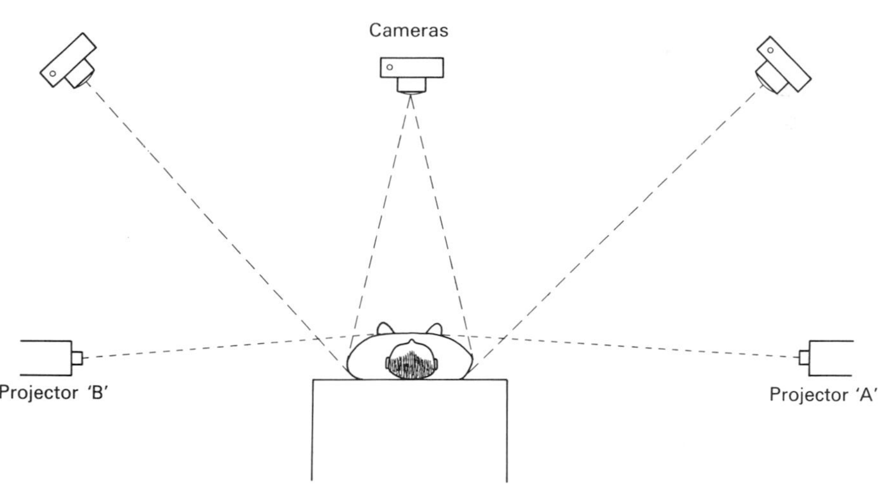

Fig. 13.1. For the recumbent patient respiratory movement occurs forwards of the firm surface. Two projectors in a horizontal plane leave a small area on the anterior chest uncovered which can be reduced rotating the patient. Two cameras which view the patient obliquely from above provide maximum coverage, but in practice a single overhead camera is sufficient.

may be worthwhile if the amount of information obtained is significantly increased. We therefore made a study of slices of the body taken with CT scans in inspiration and expiration. On top of this were drawn to scale the fields of view of one overhead camera and of a pair of cameras at 45° to the angle of projection. Movement of the surface of the body during respiration may take place within the field of view of one or two cameras or even outside the view of both. Although motion occurs outside the field of view, when it is weighted for the importance of its contribution to respiration by measuring the area change within each field compared to the overall area change, then on average 99 per cent of respiratory motion can be captured by a single overhead camera. So the apparatus has been built at Stoke Mandeville Hospital with a single overhead camera and two projectors in the horizontal plane. The patient's bed can easily be wheeled underneath and either the bed or the whole apparatus can be adjusted to a suitable height to provide an even spread of contours. With the room lights extinguished the patient is illuminated with contour lights over the trunk which change pattern as he breathes.

Ideally we would want to be able to quantify the information obtained from the video record and we intend to develop this. For the moment we rely on the single image from a 35 mm camera for measurement. A remotely controlled Olympus OM2 camera with motor drive is used and the image obtained on ordinary 400 ASA black and white film. The film is developed in the usual way and the negatives are projected on to a digitizing table under conditions of known magnification. The contours are identified by eye and each one is traced by a hand held cursor. The untransformed x coordinates are entered at y levels 10 mm apart and the z dimension is given by the contour identification. This untransformed data is stored as a data file. The data can be retrieved and the true spatial coordinates are calculated by correction for the divergent optics by simple geometry. Then the true coordinates form the data base for any subsequent program.

RESULTS AND DISCUSSION

There are two ways in which the contour lines visible on the patient can be recorded and analysed. First, the image can be obtained by a 35 mm still camera in a known position and used to obtain quantitative information about the surface of the trunk. Secondly, a video recording from a camera in a similar position gives a record of the change in contour pattern with breathing. With a little practice an observer can interpret the relative movements of the rib cage and the abdomen in different conditions (Goldman 1982). A video record may also display patterns of breathing which will change over time, such as occurs with respiratory muscle fatigue. Examples of such differences in the pattern of breathing movements are seen in subjects and patients whose breathing is maximally stimulated by rebreathing carbon dioxide. In quiet breathing in the normal supine subject the diaphragm is the major muscle of respiration whose action is reflected by displacement of the abdomen and some expansion of the lower rib cage. As ventilation is stimulated the contribution of the rib cage increases as intercostal and accessory muscles are actively recruited. At maximal stimulation the rib cage and abdomen expand almost equivalently. In the tetraplegic patient the intercostal and abdominal muscles are paralysed and the diaphragm is the major remaining muscle of respiration. Under normal circumstances it should be more than sufficient for their meagre ventilatory requirements. However, following injury their vital capacity falls to a fraction of what is expected because the unsupported rib cage collapses under the negative intrapleural pressure produced by the diaphragm's action and most of the work of breathing goes into deforming the rib cage with a resulting decrease in airflow. Patients with ankylosing spondylitis also have abnormalities of their rib cage which affect their breathing. In this case the muscles are intact but the joints of the rib cage have restricted movement and the chest is unable to expand correctly. The diaphragm again is the predominant muscle of respiration which displaces the abdomen and preferentially ventilates the lower zones of the lung, a fact which may account for a prediliction to upper lobe fibrosis. These patients are better off than the tetraplegics because although the intercostal muscles are unable to expand the rib cage, they can at least withstand negative intrapleural pressure and prevent paradoxical movement of the rib cage. The video record then allows the clinician to extend the physical examination to provide a qualitative record of the thoracoabdominal movements with respiration.

Error in the measurement of any one point is cubed in the measurement of volume and therefore we believe that the measurement of volume is a severe test of the accuracy of the system. In the original apparatus measurement of the volume of an object is easily testable since both front and back views are 'stuck' together. In the single-camera apparatus the absolute volume of an object can only be measured if its back boundary is visible. With the human torso the inability to define a posterior boundary is unimportant, we are concerned with change in shape or volume with respiration rather than absolute measurement. The best way to measure change in volume is to follow the vital capacity, the deflation of the chest with exhalation, and compare the volume change of the chest measured optically with the volume change at the mouth with a spirometer. The pictures are taken at the extremes of respiration and at approximately one litre intervals through the breath. The subject breathes out into a spirometer whose digital readout is visible in the picture (Fig. 13.2). We can then compare measured air flow with optically derived volume change. In many normal subjects and tetraplegic patients the results show good correlation, but with an error of up to 400 ml in a vital capacity of six litres. For the individual the regression lines are much tighter and the accurate recovery of volume change may depend upon chest shape.

For a more complete interpretation of the mechanics of the chest wall, its

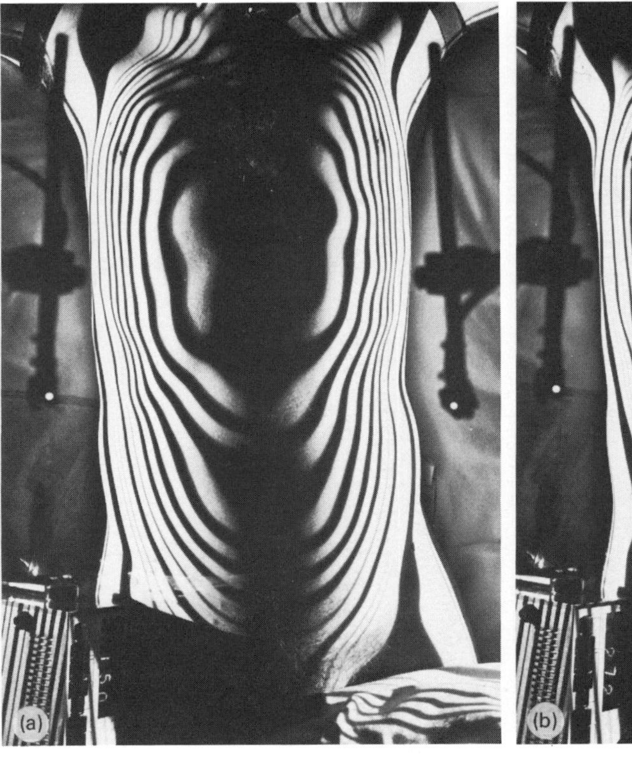

Fig. 13.2. A tetraplegic patient in inspiration (a) and expiration (b). The visible change in contour pattern is mainly abdominal. The light-emitting diodes provide the magnification factor for quantitative measurements and the actual volume change can be compared with the spirometric output visible at the bottom left.

shape and motion must be described. Techniques that are available to make these measurements such as calipers or magnetometers are uni-dimensional and the latter require frequent calibration. The use of these methods on patients who are seriously ill is often prohibited by the disturbance it produces. Optical contour mapping is ideal for making surface measurements of the chest wall in these circumstances. The information obtained can be displayed as profiles chosen at selected levels which may be visible or marked on the patient beforehand. The level profiles are then overlayed to give a picture of the movement through the breath. An example is illustrated by a patient whose breathing is supported entirely by the artificial electrical stimulation of the left phrenic nerve (Fig. 13.3). The pattern of distortion of the thoraco-abdominal wall is very complicated and the change in total volume is only 400 ml. The profiles at selected levels describe some of the motion of the rib cage and abdomen.

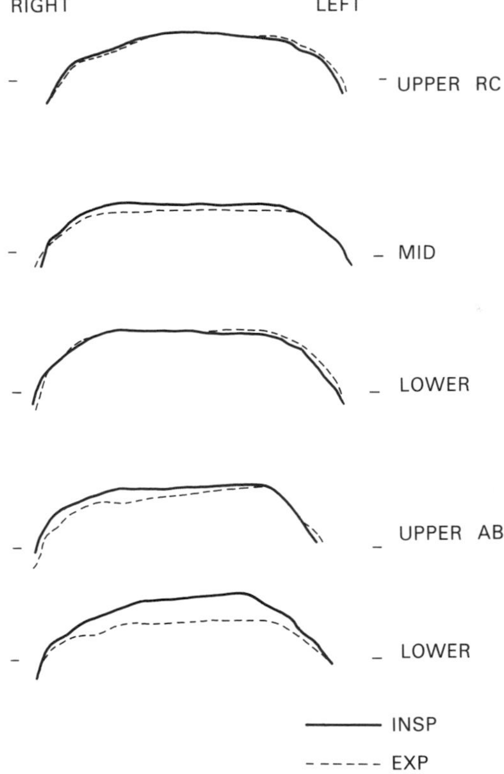

Fig. 13.3. Overlayed inspiration and expiration profiles of the anterior thoraco-abdominal wall taken at levels through the rib cage (RC) and abdomen (AB). Note the paradoxical motion of the rib cage at two levels and the asymmetry of abdominal expansion.

At the level of the diaphragm, the rib cage moves paradoxically during inspiration as does the left upper chest. The rest of the rib cage expands poorly and abdominal expansion predominates, but is asymmetrical.

CONCLUSIONS

The advantages of a technique such as optical contour mapping are obvious, since they allow us to make complex measurements of the dimensions of the thoraco-abdominal wall with minimal intervention or disturbance to the patient. This is most important in the study of those patients who have been previously inaccessible to investigation because of the nature of their illness. The limiting factor in the widespread clinical application of the method is in the digitization and analysis of the image. The individual pictures still have to be digitized by hand and each contour identified by eye, a process which currently takes about 20 minutes per picture. The development of automatic or semi-automatic contour identification to speed up this process will make its routine use worthwhile. For the present, the value of the video record of contour pattern change extends clinical observation to allow assessment of the individual patient, while the quantitative measurements are helping us to begin to explain the complex motions of the chest wall.

Acknowledgements

The collaborative study at Stoke Mandeville is supported by the Medical Research Council.

REFERENCES

Goldman, M. D. (1982). Interpretation of thoraco-abdominal movements during breathing. *Clin. Sci.* **62**, 7–11.

Morgan, M. D. L., Courlay, A. R., and Denison, D. M. (1984). Optical mapping of the thoraco-abdominal wall in the recumbent patient. *Thorax* **39**, 101–6.

Peacock, A. J., Morgan, M. D. L., Gourlay, A. R., Turton, C., and Denison, D. M. (1984). Optical mapping of the thoraco-abdominal wall. *Thorax* **39**, 93–100.

14 An Optical Technique of Mapping the Change in Shape of the Chest Wall with Breathing

A. J. PEACOCK, M. D. L. MORGAN,
A. R. GOURLAY, AND D. M. DENISON

INTRODUCTION

In principle, any physiological event that changes the conformation of the thoraco-abdominal wall can be observed and quantified to the extent to which that conformational change is an accurate reflection of the physiological process. This is true whether the change is due to processes occurring in the lung, or is due to processes occurring in the chest wall itself. This principle has not been lost on physicians specializing in diseases of the chest, and routine clinical examination of patients with respiratory disease always includes an assessment of the movements of the rib-cage and abdomen. However, respiratory function tests have not followed the lead of respiratory medicine because, while we can measure accurately the volumes of gas in the chest, the rate of inspiration and of expiration, and the rate of absorption of gas from the airspaces to the bloodstream, there is no widely available technique for measuring the shape of the thoraco-abdominal wall and the change in that shape with breathing. The problem is that the trunk is a complex shape, especially in women, and is difficult to measure. Workers have used many different methods to determine the change in dimensions of the rib-cage and abdomen with breathing, for example tape measures, calipers, magnetometers, and inductance plethysmography; but all these techniques suffer from at least two inadequacies. First, they measure only one or two dimensions of a three-dimensional (3D) structure, and secondly, they all rely on physical contact, which must cause some reactive error, no matter how small. Optical techniques avoid both these criticisms. Whether the map of thoraco-abdominal wall motion is generated by light projection, Moiré interferometry or holography, the surface of the body being measured remains untouched by the measuring instrument. Such optical techniques can provide accurate three-dimensional measurement of the change in shape of the chest wall with breathing.

METHOD

At the Brompton Hospital we have developed an optical-mapping system based on light-projection for studying breathing. We use it to study both patients with abnormal chest walls, e.g. scoliosis, and patients in whom chest-wall motion is disturbed by their underlying lung disorder, e.g. emphysema. The details of the technique are described elsewhere (Peacock *et al.* 1984) but in brief we have arranged two 35 mm cameras and two ordinary projectors around a reference frame in which the patient stands. The cameras are fired simultaneously at speeds up to 4 frames per second by remote control boxes. The projectors project 35 mm slides on which have been etched a pattern of black and white stripes. The reference frame has a platform that can be adjusted for the patient's height and markings facing the two cameras, which are used to define the optical planes of the system and the magnification factor in the photographic image. The patterns, which are projected from both sides, cover virtually the whole surface of the patient's trunk, forming contours which can be recorded by the cameras situated in front and behind. These analog records (Figs. 14.1(a) and (b)) are then encoded digitally and the digital information can be used to calculate volume, surface area, cross-sectional area at any level and plot isometric (Figs. 14.2(a) and (b)) or isocontour maps of trunk shape and the change in that shape with breathing. Using this technique we can measure trunk volume and cross-sectional area to within 5 per cent. The volume determination can be used in several ways (Peacock *et al.* 1982). For example, if photographs are taken at the beginning and the end of a breath, one can measure respired volume; if they are taken at the beginning and end of a breath-hold following the inhalation of a known concentration of a soluble gas such as nitrous oxide, one can measure trunk shrinkage which is proportional to blood flow through the lungs; if they are taken before and after blowing into a manometer to a pre-determined pressure, the compressible gas volume can be measured. To make these measurements clinically useful, the error in volume calculation must be less than 0.5 per cent—a tenfold improvement on the error with which we can measure test objects such as a shop-window mannequin. This sounds difficult, but in fact we have found that the error is not in the optics of the projection and recording system but in the reconstruction of the trunk from the recorded data. If the mannequin is simply rotated through 45° while remaining in the centre of the reference frame, the volume estimated can vary by as much as 15 per cent. If, however, there is no rigid-body motion the variation is less than 0.1 per cent. Because of the pronounced effect of rigid-body motion on the accuracy of the light-projection system the patients now sit on a specially-constructed seat which supports both the pelvis and lower back. Their shoulders are also braced so that upper and lower body are effectively but comfortably restrained without any significant loss of optical information. With this modification we have been able to measure respired volume with an error of less than 5 per cent. Since the change in volume of the trunk with breathing varies between 0 and 15 per cent of its total, the error in calculation of total volume must be less than 0.75 per cent.

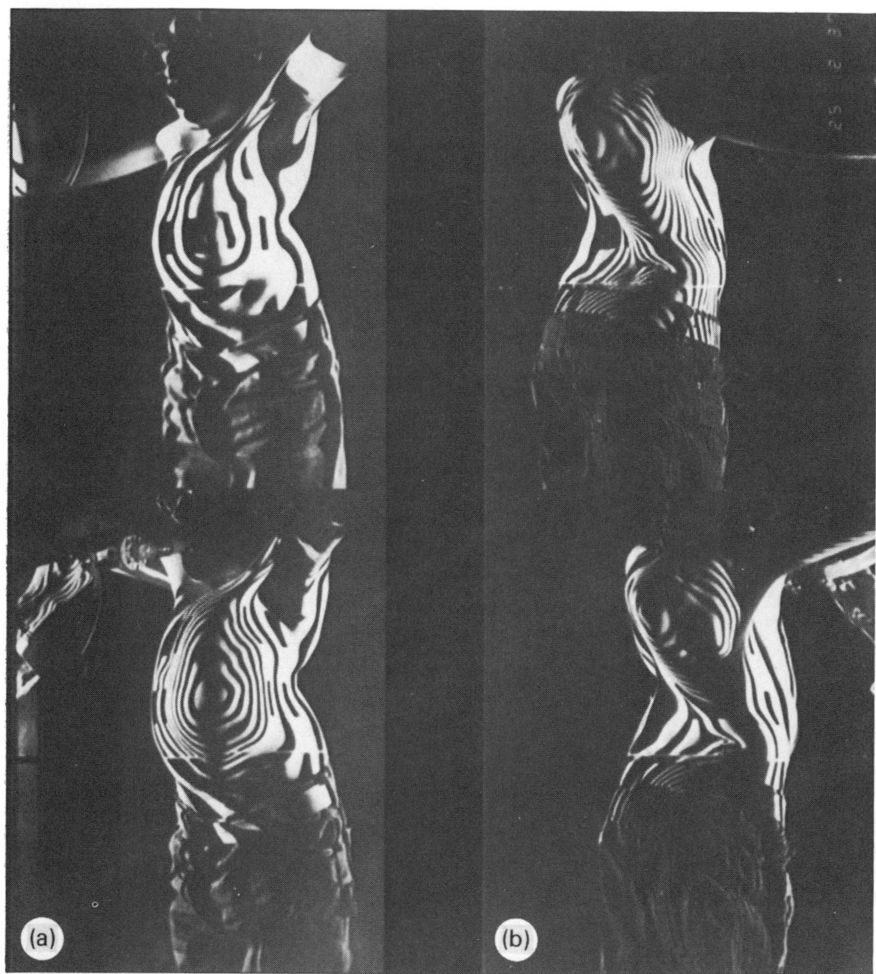

Fig. 14.1. Optical contour maps of the front (a) and back (b) of a scoliotic at residual volume (*above*); total lung capacity assisted by a constant volume (one litre) positive pressure ventilator (*below*).

CLINICAL APPLICATIONS

While it is attractive to measure such indices of respiratory function by optical contouring, these variables can be measured very adequately by other instruments, and the real clinical usefulness of the technique will be in looking at the spatial distribution of each breath, and relating the observed patterns of distribution to the underlying respiratory disorder. This approach has been applied to the study of patients with emphysema and scoliosis. In emphysema the hyperinflated lungs cause flattening of the diaphragms from their normal cupola

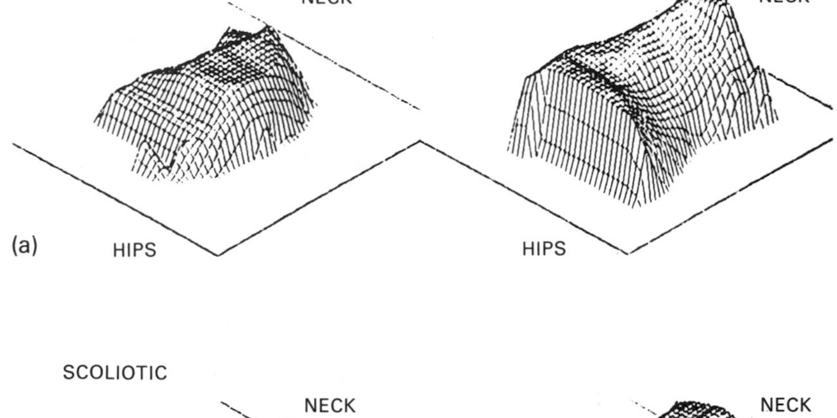

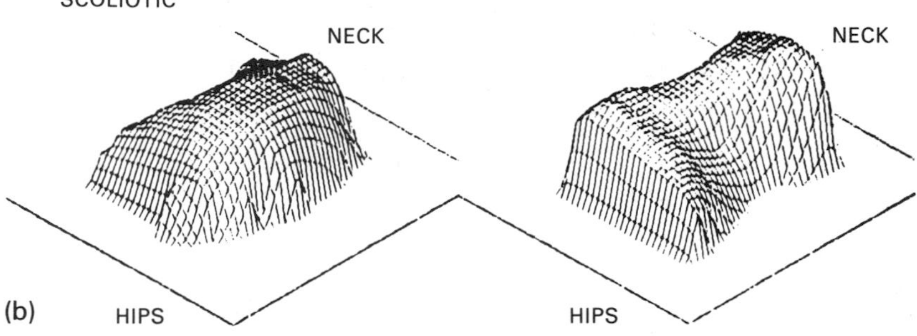

Fig. 14.2. Isometric plots of the digital information encoded from the photographs in Figs. 14.1 (a) and (b).

shape so that when a patient with this disease inspires the diaphragm causes (paradoxical) inward motion of the lower ribs. In this case the diaphragm—which is normally responsible for 60 per cent of inspiration in quiet breathing—may actually be inhibiting inspiration and a study is under way to show whether temporary paralysis of the diaphragm by injecting local anaesthetic around the phrenic nerves will improve ventilation. Optical mapping is being used to demonstrate the paradoxical motion and show any improvement following the injection. In scoliosis the distorted spine and rib-cage result in a decrease in chest wall compliance which in turn increases the work of breathing and decreases the amount of air that can be moved in and out (Bergofsky 1979). This effect can be clearly seen in the optical contour maps and isometric plots of the front and back of a patient with severe scoliosis, as he inspires from residual volume to total lung capacity, with the help of a ventilator (Figs. 14.1 and 14.2). The total volume inspired is only one litre, compared with about five litres for a healthy adult male, and this volume change is occurring almost entirely in the abdomen, because the rib-cage is fixed and immovable. The back takes no part in respiration but is seen to twist as the patient inspires.

CONCLUSIONS

The two examples, patients with a primary chest wall disorder and patients in whom the disordered chest wall motion is secondary to underlying lung disease, demonstrate the usefulness to respiratory medicine of a three-dimensional measuring system such as optical mapping. The mechanical defect can be clearly and objectively demonstrated in a totally non-invasive way using cheap equipment. These advantages allow repeated measurements, which can be used to record the progress of the disease and, of course, show any improvement with treatment.

REFERENCES

Bergofsky, E. H. (1979). Respiratory failure in disorders of the thoracic cage. *Am. Rev. resp. Dis.* **119**, 643–8.

Peacock, A. J., Morgan, M. D. L., Gourlay, A. R., and Denison, D. (1982). Non-invasive assessment of several respiratory functions at the same time. *Am. Rev. resp. Dis.* **125**, 216.

—, —, —, Turton, C., and Denison, D. M. (1984). Optical mapping of the thoraco-abdominal wall. *Thorax* **39**, 93–100.

15 Back Surface Shape Relationship to Spinal Deformity

I. A. F. STOKES, L. C. COBB, M. H. POPE, AND M. S. MORELAND

INTRODUCTION

The clinical assessment of scoliosis remains a major orthopaedic challenge. The clinical management depends crucially upon the detection of progression and, if treatment is initiated, it is valuable to have again a sensitive technique for detection of change. Orthopaedic management tends to focus upon the spinal deformity, whereas the surface deformity and its cosmetic implications may be of greater interest to the patient. In a 'preclinical' sense, the geometric assessment of scoliosis is becoming important as programmes are developed for the screening of schoolchildren. Here the challenge is primarily to detect all asymmetries, and secondarily to reject those which are benign and not likely to lead to progressive deformity.

New techniques for studying the geometry of the scoliotic deformity have been developed recently. Moiré topography, which can document the back surface shape, was used by Adair *et al.* (1978) in a school screening programme and Laulund *et al.* (1982) have documented good sensitivity and specificity for the technique. This technique (Moreland *et al.* 1981*b*) is becoming extensively used in clinical documentation, although the exact relationship between the back surface shape (as shown by the moiré fringe) and the skeletal deformity (as seen on radiograph) is still not well defined (Willner 1979, 1981; Harada *et al.* 1981; Moreland *et al.* 1981*a*; Shinoto 1981). Biplanar radiography and stereoradiography have been used in clinical studies for accurate three-dimensional description of the deformity (Kratky 1975; Brown *et al.* 1976; Gibson *et al.* 1978; Raso *et al.* 1980; DeSmet *et al.* 1980; Allard *et al.* 1980; Hindmarsh *et al.* 1980; Reuben *et al.* 1982). Computerized tomography as a radiographic technique can show the deformity in serial cross-sections, including both the vertebral alignment and the rib and trunk outline (Aaro and Dahlborn 1981*a,b*).

All of these new techniques have helped to document the highly complex geometrical nature of the deformity in scoliosis. The lateral deviation of vertebrae, axial rotation of vertebrae, deformity of the rib cage, and possibly alteration of the lordosis and kyphosis, all combine to produce the cosmetic deformity and its associated disability. Lovett (1905) attempted to understand these interactions better by comparison with the normal kinematics of the healthy spine and trunk. He did this without the help of radiography which was later available to Arkin (1950) who continued with this approach. White (1969) and Lysell (1969) used biplanar radiographic techniques to study segmental mechanics in excised spines in three dimensions. They were also especially interested in the interaction between lateral bending and axial rotation in these specimens. Arkin (1950) had found differences in this behaviour depending on the state of flexion or extension of the spine, and Somerville (1952) and Roaf (1958) pursued this idea further in theories about the mechanism of scoliosis, and its relationship to lordosis and kyphosis. The axial rotation component of the deformity is especially interesting and has been difficult to document in patients because of measurement problems. Nash and Moe (1969) and Mehta (1973) have proposed methods of measuring vertebral rotation based on pedicle offset but the accuracy problems associated with these techniques have been graphically demonstrated by Coetsier *et al.* (1977) and Benson *et al.* (1976).

In this study, we have used sensitive three-dimensional measurement techniques to document the shape of both the spine and the back surface shape in a group of patients with adolescent idiopathic scoliosis. Previously (Moreland *et al.* 1981*a*), we had examined a population of 950 patients, attempting to find relationships between measurements of the asymmetry of moiré fringe photographs and the value of the Cobb angle. No good predictive relationships were found. The best correlation was among measurements of 95 patients with mid-thoracic scolioses where a correlation coefficient of 0.71 was found. A number of characteristic patterns in the moiré fringes were noted which were associated with the level, size, and magnitude of each scoliotic curve. A randomized, blind trial involving five observers was devised in order to assess the capability of this pattern recognition scheme for diagnosing curve types based on the appearance of the moiré fringe photographs (Moreland *et al.* 1983). This study demonstrates that high predictive values could be achieved for assessing the presence, the side and the anatomical region of a scoliotic deformity in a moiré fringe photograph alone, but the ability to assess the magnitude of the deformity was considerably less accurate. These findings highlighted the need for better measurement of spine shape in three dimensions, as has been used in the present study.

METHODS

Patients in the scoliosis clinic for whom a clinical, frontal plane radiograph was indicated, were asked to participate in this special study. The measurement protocol involved stereoradiography of the entire spine and quasi simultaneous back surface topography. Patients would stand in front of an X-ray film holder while a stereo-pair of low-dose radiographs (technique of Ardran *et al.* 1980) was made on separate films using a PA tube position and an additional tube giving a 15° oblique exposure on a second film. Immediately afterwards, a Fujinon FM–40 projection moiré camera was moved into position immediately behind the patient. This provided a moiré fringe photograph and additionally the camera had been modified to project also a raster or grating of lines at approximately 20 mm spacing on to the patient's back. This raster pattern was photographed from a different angle to provide a readily digitizable record of the patient's back surface (Fig. 15.1). The apparent distortion of the grating when

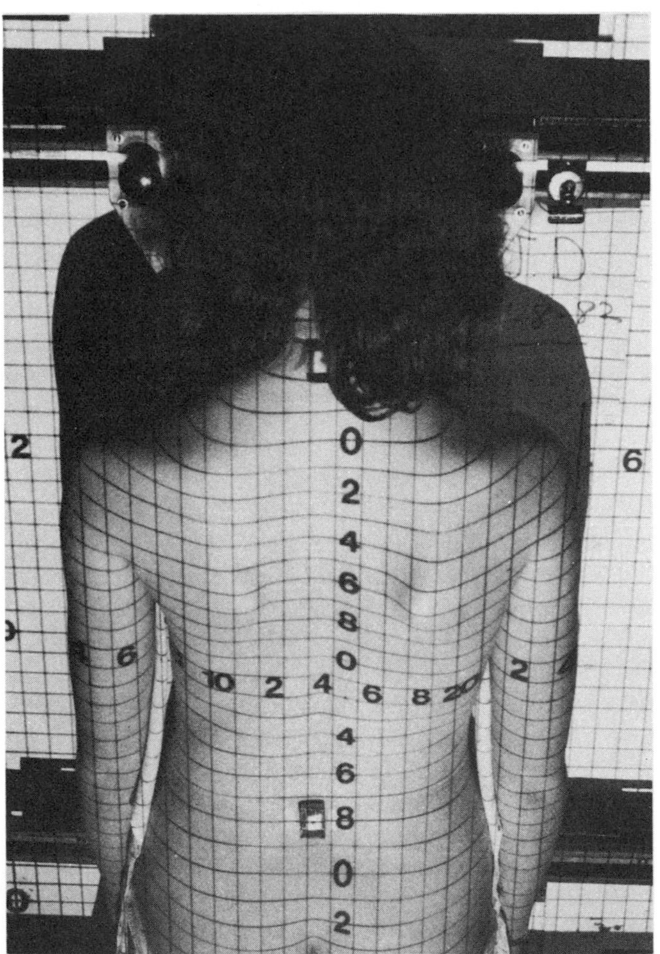

Fig. 15.1. Example of a raster photograph used for documenting back surface shape. The apparent distortion of the square grid of lines projected optically on to the patient's back contains parallax information for the reconstruction of the back surface shape.

viewed from the different angle contained parallax information for reconstruction of the three-dimensional coordinates of images points as described by Frobin and Hierholzer (1982). A computerized reconstruction of the spine shape was made from the stereoradiographs. Six anatomic landmarks were marked on the image of each vertebra in each of the pair of radiographs. The six landmarks were: upper and lower endplate centres and the upper and lower margin of the base of each pedicle. The position and orientation of each vertebral body was defined by these landmarks which are on the vertebral centrum and derived measurements were independent of any distortion of the posterior elements. The same test object was used for calibration of both the radiograph and raster techniques, so that the three-dimensional coordinates of points on the back surface and the three-dimensional coordinates of spinal landmarks were all in the same Cartesian coordinate set.

Details of the first 36 patients who have been studied by these techniques are shown in Table 15.1. Overall, there were 56 structural curves among these

Table 15.1. *Details of 56 structural curves in 36 patients studied by means of stereo radiography and raster topography. For the purposes of this study, curves were divided into two regions, lumbar and thoracic, with thoraco-lumbar curves classified as lumbar*

	No. of curves	Cobb angle	
		Range	Mean
Overall	56	5–59°	22.13°
Thoracic (T1–T10)	27	6–58°	22.41°
Lumbar (T11–L5)	29	5–59°	21.86°

patients, and there was a wide range of Cobb angle with a mean of 22°. For the purposes of this study, the curves were divided into two regions, thoracic and lumbar, with roughly equal numbers of curves in each group. Thoracic curves were those with an apex at T10 or above, and lumbar curves were those with an apex at T11 or below. Thus the curves with an apex in the thoracolumbar region were not considered separately but included with the lumbar curves.

The patients have been studied by examining a computer reconstruction of a cross-section on a horizontal plane through the patient at the level of each vertebra from thoracic 4 to lumbar 5, as shown in Fig. 15.2. The asymmetry of the back surface was characterized by drawing a double tangent across the two maxima of the cross-section and measuring the rotation of this line relative to the frontal plane. In the cross-sections shown in Fig. 15.2, this angle is the angle between the double tangent and the horizontal. The orientation and position of each vertebra was characterized by measuring from the anatomic

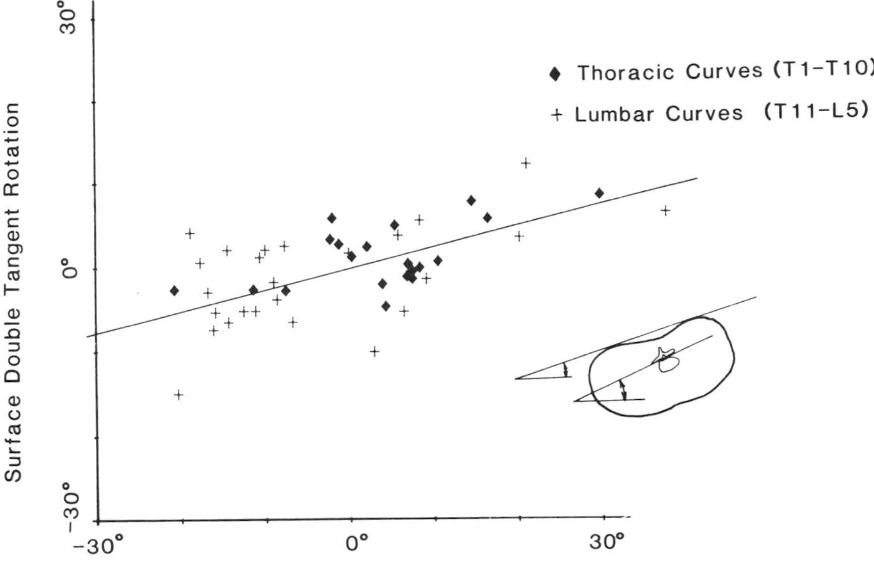

Fig. 15.4. Correlation between the amount of back surface rotation, (represented by the amount of surface double tangent rotation from the frontal plane) and the amount of vertebral axial rotation. It can be seen that for nearly all the curves, the amount of back surface rotation is considerably less than the amount of vertebral rotation, since the gradient of the regression line is shallow.

level in the curve at which this occurred. In this case the coefficient of correlation was 0.75. This reflects differences between thoracic and lumbar curves. In the thoracic curve the back surface rotation reaches its maximum value at a level lower than the curve apex. The correlation coefficient was even higher (0.85) if only patients with single structural curves were considered in the analysis. This is evidence of an interaction between curves when there are multiple structural curves in the same patient.

DISCUSSION

The relationships between the various components of the scoliotic deformity in these patients have emphasized the large variability between patients. In general, the measurement reported here did not give significantly different findings between the thoracic and lumbar curve groups. This apparently de-emphasizes the rib cage as a structure which would produce an inherently different scoliosis in the thoracic and lumbar regions. An exception to this is the findings for the level of the maximum surface rotation which occurs lower than the apex of a curve in the thoracic spine.

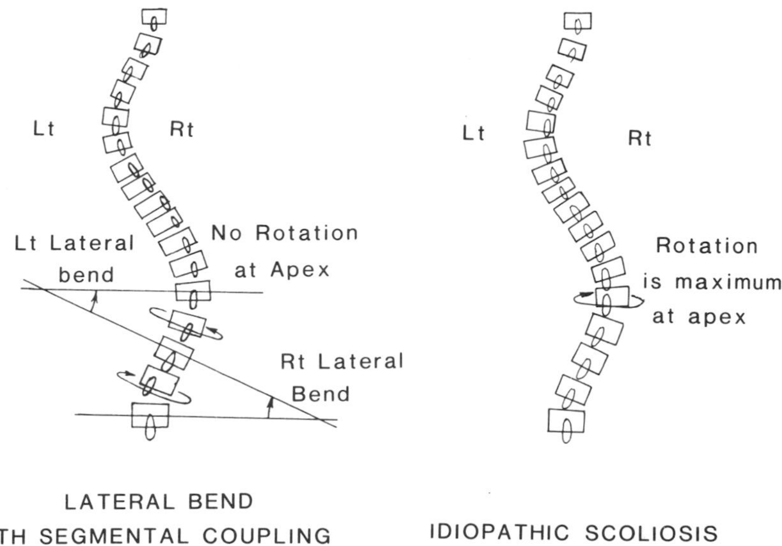

Fig. 15.5. Theoretical analysis of the implications of coupling mechanisms in the spine as they relate to the scoliotic deformity. On the left, the process of bending a straight, normal spine into a scoliosis is illustrated. It is assumed that the intervertebral joints have a mechanism by which axial rotation is produced in an amount proportional to the amount of lateral bending in these joints, but in the opposite sense. This mechanism would not produce the pattern of rotation seen in idiopathic scoliosis, as illustrated on the right.

The development of the axial rotation of vertebrae at the apex of a scoliotic curve has often been seen as the result of a mechanism produced by segmental coupling of the lateral bending motion with the axial rotation in the intervertebral joints. This implies that scoliosis is primarily a lateral deviation asymmetry, but that some inherent mechanism in intervertebral joints (motion segments) of the spine produces the accompanying axial rotation. An argument based on symmetry, illustrated in Fig. 15.5, shows that this explanation is not satisfactory. On the left of this figure, the process of bending a straight, normal spine into a scoliosis is illustrated. Starting from the lowest end, the spine is required to bend first to the right and then to the left in order to reach the apex. If right lateral bending produced left axial rotation and left bending produced axial rotation to the right, then it would follow that there would be no net axial rotation at the apex of the curve. In other words, the rotation produced by the right bending would be cancelled out by rotation in the opposite sense produced by the left bending. This contrasts with the actual relationships seen in idiopathic scoliosis, illustrated on the right of Fig. 15.5. We have concluded from this theoretical analysis that this form of 'segmental coupling' could not be responsible for the generation of the axial rotation within these curves. The implication is that the mechanical interactions within the spine which produced the multiple-dimensioned deformity must be interactions over longer segments of the spine than are represented by two vertebrae. This suggests that the structures such as ligaments and muscles which cross more than one segment of the spine are involved intimately in the generation of the deformity.

We have found that the relationship between the deformity of the back surface shape and the skeletal deformity is rather elusive. This has been found in previous attempts to relate the moiré ringe asymmetry to the Cobb angle, and in a study of rib hump and scoliosis reported by Thulbourne and Gillespie (1976). One factor we have found to help explain the variation is the difference between lumbar and thoracic curves in the levels at which the back surface deformity becomes maximal. However, the magnitude of the back surface rotation was found to be very similar for curves of the same magnitude in the different regions. It was surprising to us that the axial rotation of vertebrae apparently contributed so little to the back surface rotation, according to the correlation we obtained.

In future work we intend to gain further insight into these complex interactions by means of multiple regression analysis, and by means of mathematical modeling of the mechanisms of spine and rib cage deformations. We hope this will also give a better understanding of the categorization of these deformities by the anatomic region and their back surface expression.

CONCLUSIONS

1. In three dimensions there are complex inter-relationships between surface and skeletal measures of the scoliotic deformity. For clinical documentation, the Cobb angle alone is an incomplete measure of the extent of the deformity.
2. Coupling of axial rotation to lateral bending in intervertebral segments would not produce a scoliosis with the correct amount of axial rotation, according to our theoretical analysis and experimental findings.
3. The magnitude of the axial rotation of vertebrae in a scoliotic curve is related to the Cobb angle, but the correlation is not strong. The amount of rotation is apparently not different in thoracic and lumbar curves of the same severity.
4. The amount of back surface deformity associated with a scoliotic curve is more closely correlated with the amount of lateral deviation in the curve than it is correlated with the amount of vertebral axial rotation.

Acknowledgements

This work was done with financial support of NIH Grant #RO1 AM 30235, and with equipment provided by the Fuji Photo Optical Company.

REFERENCES

Aaro, S. and Dahlborn, M. (1981*a*). Estimation of vertebral rotation and the spinal and rib cage deformity in scoliosis by computer tomography. *Spine* **6**, 460–7.

— — (1981*b*). The longitudinal axis, rotation of the apical vertebra, the vertebral, spinal and rib cage deformity in idiopathic scoliosis studied by computer tomography. *Spine* **6**, 567–72.

Adair, I. V., van Wijk, M. C., and Armstrong, G. W. D. (1978). Moiré topography in scoliosis screening. *Clin. Orthopaed.* **129**, 165–71.

Allard, P., Duhaime, M., Raso, J. V., Thiry, P. S., Drouin, G., and Geoffroy, G. (1980). Pathomechanics and management of scoliosis in Friedreich ataxia patients. Preliminary report. *Can. J. neurol. Sci.* **7**, 383–8.

Ardran, G. M., Coates, R., Dickson, R. A., Dixon-Brown, A., and Harding, F. M. (1980). Assessments of scoliosis in children: low dose radiographic technique. *Br. J. Radiol.* **53**, 146–7.

Arkin, A. (1950). The mechanism of rotation in combination with lateral deviation in the normal spine. *J. Bone Jt Surg.* **32A**, 180–8.

Benson, D. R., Schultz, A. B., and DeWald, R. L. (1976). Roentgenographic evaluation of vertebral rotation. *J. Bone Jt Surg.* **58A**, 1125–9.

Brown, R. H., Burstein, A. H., Nash, C. L., and Schock, C. C. (1976). Spinal analysis using a three-dimensional radiograph technique. *J. Biomech.* **9**, 355–65.

Coetsier, M., Vercauteren, M., and Moerman, P. (1977). A new radiographic method for measuring vertebral rotation in scoliosis. *Acta orthopaed. belg.* **43**, 598–605.

Deane, G. and Duthie, R. B. (1973). A new projectional look at articulated scoliotic spines. *Acta orthopaed. scand.* **44**, 351–65.

DeSmet, A. A., Tarlton, M. A., Cook, L. T., Fritz, S. L., and Dwyer, S. J. (1980). A radiographic method for three-dimensional analysis of spinal configuration. *Radiology* **137**, 343–8.

Frobin, W. and Hierholzer, E. (1982). Analysis of human back shape using surface curvatures. *J. Biomech.* **15**, 379–90.

Gibson, D. A., Koreska, J., Robertson, D., Kahn, A., and Albisser, A. M. (1978). The management of spinal deformity in Duchenne's muscular dystrophy. *Orthopaed. Clins N. Am* **9**, 437–50.

Harada, Y., Takemitsu, Y., and Imai, M. (1981). The role of contour line photography using light cutting method and Moiré photography in school screening for scoliosis. In *Moiré fringe topography and spinal deformity* (ed. M. S. Moreland, M. H. Pope, and G. W. D. Armstrong) pp. 113–21. Pergamon Press, New York.

Hindmarsh, J., Larsson, J., and Mattsson, O. (1980). Analysis of changes in the scoliotic spine using a three-dimensional radiographic technique. *J. Biomech.* **13**, 279–90.

Kratky, V. (1975). Analytical X-ray photogrammetry in scoliosis. *Photogrammetria* **31**, 195–210.

Laulund, T., Sojbjerg, J. O., and Horlyck, E. (1982). Moiré topography in school screening for structural scoliosis. *Acta orthopaed. scand.* **53**, 765–8.

Lovett, R. W. (1905). The mechanism of the normal spine and its relation to scoliosis. *Boston med. surg. J.* **153**, 349–58.

Lysell, E. (1969). Motion in the cervical spine. *Acta orthopaed. scand.* Supp. 123.

Mehta, M. H. (1973). The radiographic estimation of vertebral rotation in scoliosis. *J. Bone Jt Surg.* **55B**, 513–20.

Moreland, M. S., Barce, C., and Pope, M. H. (1981*a*). Moiré topography in scoliosis: pattern recognition and analysis. In *Moiré fringe topography and spinal deformity* (ed. M. S. Moreland, M. H. Pope, and G. W. D. Armstrong) pp. 171–85. Pergamon Press, New York.

—, Pope, M. H., and Armstrong, G. W. D. (eds.) (1981*b*). *Moiré fringe topography and spinal deformity*. Pergamon Press, New York.

—, —, Stokes, I. A. F., and Weierman, R. J. (1983). Diagnostic uses of Moiré topograms in scoliosis. *Proceedings of the 50th Annual Meeting of American Academy of Orthopaedic Surgeons*, Anaheim.

Nash, C. L. and Moe, J. H. (1969). Study of vertebral rotation. *J. Bone Jt Surg.* **51A**, 223–9.

Raso, J., Gillespie, R., and McNeice, G. (1980). Determination of the plane of maximum deformity in idiopathic scoliosis. *Orthopaed. Trans.* **4**, 23.

Reuben, J. D., Brown, R. H., Nash, C. L., and Brower, E. M. (1982). *In vivo* effects of axial loading on double-curve scoliotic spines. *Spine* **7**, 440–7.

Roaf, R. (1958). Rotation movements of the spine with special reference to scoliosis. *J. Bone Jt Surg.* **40B**, 312–32.

Shinoto, A. (1981). Quantitative analysis of scoliotic deformity by Moiré method. *J. Jap. orthopaed. Ass.* **55**, 1703–18.

Somerville, E. (1952). Rotational lordosis: the development of the single curve. *J. Bone Jt Surg.* **34B**, 421–7.

Stokes, I. A. F., Moreland, M. S., Pope, M. H., Cobb, L. C., and Armstrong, J. G. (1983). Correlation of back surface topography with 3–D spine shape in scoliosis. In *Moiré fringe topography and spinal deformity*, Vol. 2. *Proceedings of the 2nd International Conference* (ed. B. Drerup, W. Frobin, and E. Hierholzer). Fischer, Stuttgart.

Thulbourne, T. and Gillespie, R. (1976). The rib hump in idiopathic scoliosis. *J. Bone Jt Surg.* **58B**, 64–71.

White, A. A. (1969). Analysis of the mechanics of the thoracic spine in man. *Acta orthopaed. scand.* Suppl. 127.

Willner, S. (1979). Moiré topography for the diagnosis and documentation of scoliosis. *Acta orthopaed. scand.* **50**, 295–302.

— (1981). Comparison between moiré and X-ray findings in structural scoliosis In *Moiré fringe topography and spinal deformity* (ed. M. S. Moreland, M. H. Pope, and G. W. D. Armstrong) pp. 157–65. Pergamon Press, New York.

16 Automatic Measurement of Body Shape by Means of Raster Stereography

G. J. DOCTER AND J. ENSINK

INTRODUCTION

One of the diagnostic tools serving a physician is evaluation of the shape and shape change of the human body or parts of it. In order to improve the reliability of this kind of observation, there have been made many attempts to quantify the shape of the human body. As early as 1887 Schulthess designed a mechanical drawing device that recorded the three-dimensional shape of the human back. More recently the use of moiré fringe topography has been introduced (Takasaki 1973). This technique is based on the optical interference of two line patterns and produces a contour line representation of the human body. Back shape asymmetries are then quantified as the difference in fringe numbers between corresponding points on both sides of the back (Willner 1979). This method has the advantage that asymmetries are detected directly from the photographic image of the back without the use of any computational processing of the information. In screening programmes aimed at the early detection of scoliosis this technique has proved to be successful in detecting cases of structural scoliosis. At the same time however a considerable number of false-positive cases were detected and there are contradictory reports concerning the correlation between moiré findings and radiographic analysis of the spine (Laulund *et al.* 1982). Information about the form of the spine will require a more comprehensive three-dimensional description of the human back surface. Routine clinical use of such descriptions additionally requires rapid availability of the results. Digital processing of the large amounts of data then become inevitable. Moiré topographs, however, are not most suited for automatic image analysis. The main reason is the variability of the fringe patterns due to changes in body position. In that case the so-called 'raster stereography' method can be used fruitfully.

THE PRINCIPLE OF RASTER STEREOGRAPHY

Raster stereography, also known as monoscopic photogrammetry, is a modification of standard photogrammetric procedures. The method we employ is derived from studies done at the Orthopaedic Clinic of the University of Münster, West Germany (Frobin and Hierholzer 1981).

The principle of raster stereography is illustrated in Fig. 16.1. Compared to the usual photogrammetric set-up one of the cameras has been replaced by a projector, producing a line grid pattern on the object. The original line grid has a known geometry and serves as the first image (Fig. 16.1 (b)). Seen from a different angle the parallel lines become deformed. This deformed line pattern serves as the second image (Fig. 16.1 (c)).

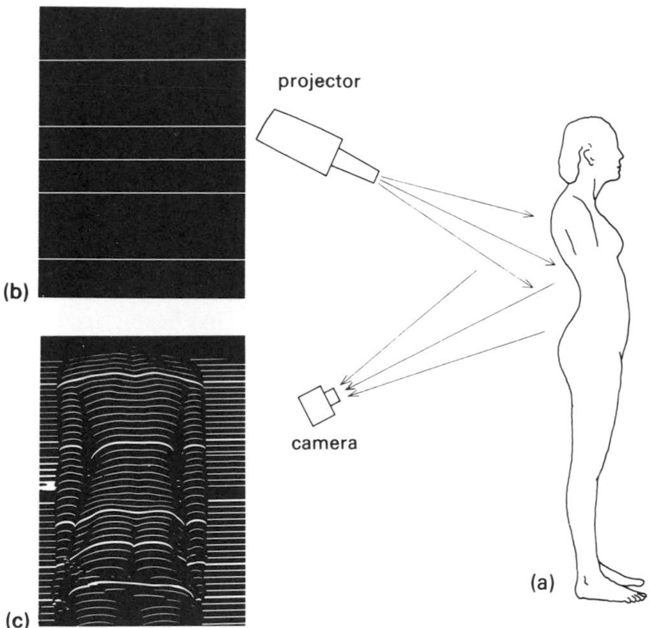

Fig. 16.1. Principle of raster stereography. (a) Measurement arrangement. (b) Projected image: coded line-grid. (c) Camera image: deformed line-grid.

Each point on an object may be represented by two stereo images, from which the original location can be reconstructed. The main feature of raster stereography is that a single image contains information from two stereo images. Each object point is represented by its coordinates in the camera image, but in the same image the location with respect to the line grid, and thus with respect to the known geometry of the projected image, can also be detected.

IMAGE ANALYSIS

The three-dimensional reconstruction follows from an analysis of the camera image. A number of steps can be distinguished in the process of detection of the deformed line-grid.

Data acquisition

A first step in analysis is to digitize the camera image and to store it for future processing. We have chosen a fully electronic digitization of the image by means of a solid-state camera (General Electric TN 2500). This camera contains a charge injection device (CID) image sensor, consisting of an array of 244 × 248 light-sensitive pixels. As compared to other image-sensing systems it has the advantage of excellent geometric linearity at the cost of resolution, although current technological developments can solve the latter problem considerably.

We interfaced this camera to an external memory (64 Kbyte static RAM) which is capable of storing a single image (Fig. 16.2). Memory-addressing is derived from the camera clock frequency. The external memory was found to be necessary because of the high video frequency of the camera, which at the other hand enables visual inspection of the image on a television monitor. The image is fed into a microcomputer, which also controls both camera and external memory.

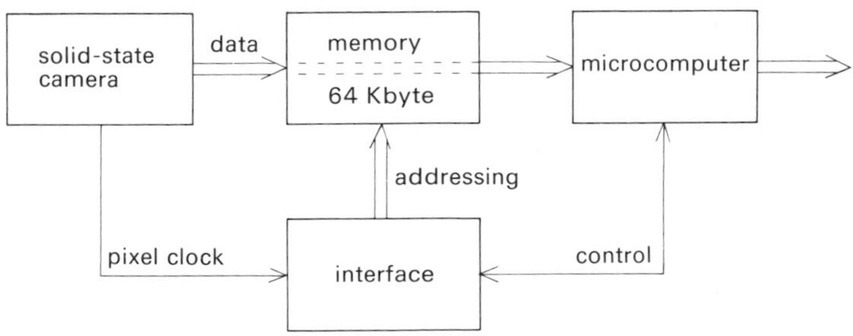

Fig. 16.2. Data-acquisition system.

Image reconstruction

The image reconstruction is based on the assumption that the raster lines are approximately perpendicular to the scanning line direction of the camera.

First, in every camera-line locations of the projected raster lines are calculated from the pixel data using a number of criteria, such as minimum light intensity levels and minimum intensity changes on the edges of raster lines. The use of interpolation techniques approximating the light intensities improves the accuracy of the raster line locations to approximately one-third of the geometric resolution. The second step in the image reconstruction is formed by reconstruction of raster line segments from neighbouring scan-lines. The reconstruction is carried out assuming a regularly shaped object with no abrupt changes in raster line location.

The third stage of the reconstruction process deals with the search for relations between adjacent line segments. Through this sequencing process the line segments are brought into a unique relative order.

Finally the correspondence between projected lines and reconstructed lines has to be determined. This absolute line numeration is accomplished by means of a line grid coding composed of a pattern of light and heavy lines (Fig. 16.1 (a)), which can be detected in the camera image. The image reconstruction results in a deformed line grid which has now been related to the originally projected line grid.

Three-dimensional reconstruction

The location of a set of control points relative to the projected line grid can be used in calibration of both the projector and the camera. We used a modification of the so-called 'bundle-method' (Frobin and Hierholzer 1982*a*). Using a redundant set of control points, variation of both image coordinates and control point locations results in an optimal estimation of camera and projector orientation.

This calibration procedure subsequently allows reconstruction of the three-dimensional shape of the object from the deformed line grid pattern of the camera image. The resulting description of the object comprises a set of three-dimensional coordinates of points distributed over the surface of the object.

DISCUSSION AND CONCLUSIONS

The objective of our study was to obtain a three-dimensional description of the human body in order to characterize normal and abnormal body shape. This has been realized by means of automatic detection and analysis of a raster stereographic image.

Presently it takes approximately 20 minutes computing time before the three-dimensional reconstruction has been obtained. It is however possible to reduce this time considerably by optimizing the measurement technique as well as the computing procedures. Accuracy studies indicate a depth resolution of approximately 2.0 mm, depending on the orientation of the surface with respect to the projector and the camera.

Preliminary clinical trials using the raster stereographic technique are also promising: the equipment is easy to operate and there is no expertise required in positioning the patients. Figure 16.3 shows an example of the kind of results that can be obtained. In this case the asymmetry of the lumbar spine is clearly shown in the horizontal cross-sections of Fig. 16.3 (b). This asymmetry can be quantified, e.g. by means of calculating correlation coefficients between both body halves. Also other methods of characterizing the human back shape have been reported (Neugebauer and Windischbauer 1982). The so-called invariant parameter description (Fobin and Hierholzer 1982*b*) seems to have particular application.

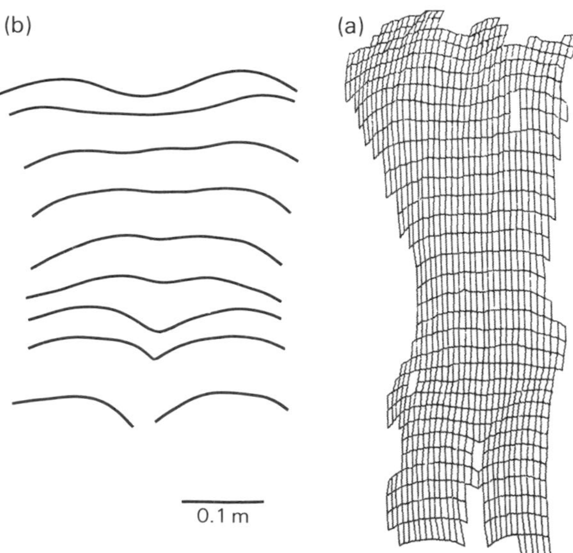

Fig. 16.3. A case of lumbar scoliosis. (a) Reconstructed surface. (b) Horizontal cross-sections.

Raster stereography has proved to be a very versatile method of shape measurement. By simple modifications of the optical arrangement other possibilities arise. Practical applications, e.g. shape analysis in fitting procedures of orthotic appliances, but also more fundamental studies, e.g. the shape of articular surfaces, become feasible with the help of raster stereography.

REFERENCES

Frobin, W. and Hierholzer, E. (1981). Raster stereography: a photogrammetric method of measurement of body surfaces. *Photograph. Engng Remote Sensing* **47**, 1717.

— — (1982*a*). Calibration and model reconstruction in analytical close-range stereophotogrammetry. Part I: mathematical fundamentals. *Photograph. Engng Remote Sensing* **48**, 67.

— — (1982*b*). Analysis of human back shape using surface curvatures. *J. Biomech.* **15**, 379.

Laulund, T., Søjberg, J. O., and Hørlyck, E. (1982). Moiré topography in school screening for structural scoliosis. *Acta orthopaed. scand.* **53**, 765.

Neugebauer, H. and Windischbauer, G. (1982). Moiré topography in scoliosis research. In *Optics in biomedical sciences* (ed. G. von Bally and P. Greguss). Springer, Berlin.

Schulthess, W. (1887). Ein Mess- und Zeichnungsapparat für Rückgratsverkrümmungen. *Zbl. Orthopaed. Chir.* **4**, 102.

Takasaki, H. (1973). Moiré topography. *Appl. Optics* **12**, 845.

Willner, S. (1979). Moiré topography for the diagnosis and documentation of scoliosis. *Acta orthopaed. scand.* **50**, 295.

Section 3:
Soft Tissue and Spine

17 Measurement of Physical Parameters at the Patient–Support Interface

M. W. FERGUSON-PELL, N. P. REDDY,
S. F. C. STEWART, V. PALMIERI,
AND G. V. B. COCHRAN

INTRODUCTION

Prevention of pressure sores continues to be one of the primary problems confronting rehabilitation specialists. One vital aspect of this subject concerns the design, selection, and evaluation of wheelchair cushions to reduce tissue trauma in the seated patient. A large number of types and models of commercially available wheelchair cushions and other seating devices exist, and many claims are made for their attributes. Despite the extensive literature on this subject, the fact remains that no comprehensive programme for evaluation of available cushions exists. Development of test methods, and eventually standards of cushions, would permit institutions and large health care organizations to make informed selections for volume purchases of cushions as well as to improve designs and facilitate prescription of specific cushions for individual patients.

Suitable test protocols require that those factors thought to be responsible for the causation of pressure sores be measured, to enable the properties of candidate cushions to be compared. Accordingly, testing techniques have been developed at Helen Hayes Hospital, in collaboration with the Veterans Administration, to compare the performance of wheelchair chusions. Test protocols were developed and intended to be applicable to as wide a variety of finished types of cushions as possible, although all of the tests could not be relevant in certain special cases, such as custom-moulded or mechanically active seating devices.

The protocol that resulted consisted of a number of bench tests devised empirically to highlight various aspects of cushion performance, measures of environmental factors at the patient–cushion interface and certain clinical factors. The results of this type of testing do not permit ranking of cushions in order of effectiveness (Cochran and Palmieri 1980) but are useful in making preliminary selections and design improvements prior to clinical trials, and can draw the attention of manufacturers to serious defects in their cushions. During this study, attention was drawn to the difficulty of measuring environmental factors at the skin–cushion interface. Before more comprehensive technical descriptions and standards for materials and cushions can be established, devices used to evaluate them require to be tested for their accuracy and reproducibility both in the laboratory and in clinical field trials. In this chapter, the performance of some widely used pressure, temperature, and humidity sensors will be reviewed.

RELEVANT ASPECTS OF PRESSURE SORE AETIOLOGY

The aetiology of pressure sores has been widely discussed (see for example, Kosiak 1959; Bailey 1967; Fernie 1973; Berecek 1975; Guttmann 1976; Torrance 1981). From these studies, it is generally agreed that ischaemia produced by prolonged application of pressure to tissues overlying bony prominences is responsible for many of the pressure sores that are observed clinically. Muscle has been found to be particularly vulnerable to such damage (Groth 1942; Hussain 1953; Kosiak 1959, 1961; Daniel *et al.* 1981). The skin, on the other hand, is frequently exposed to abrasive damage, often in the presence of moisture from urine or sweat. Such persistent maceration often results in superficial damage to skin and is a focus for infection.

Elevated skin temperature is thought to increase the metabolic demands of tissues (Brown and Brengelmann 1965; Schell and Wolcott 1976) thereby increasing their vulnerability to trauma if nutrient supply and metabolite removal is inhibited by ischaemia.

An effective programme of pressure sore prevention through the provision of improved support surfaces should reduce the effects of the following parameters:

Mechanical stresses in tissues overlying bony prominences
Elevated skin temperature
Moisture at skin–support interface
Abrasion
Immobility

Good general health, balanced nutrition, and hygiene are also of considerable importance in reducing the vulnerability of patients to pressure sores.

SELECTION AND EVALUATION OF INTERFACE PRESSURE SENSORS

Uniaxial compressive stress and shear can both be of sufficient magnitude in the sitting posture to arrest blood flow in the tissues overlying the ischial tuberosities and, in certain postures, over the coccyx and sacrum. A wide range of pressure-measuring devices has been reported in the literature for monitoring the pressure at the interface between support surfaces and the skin. These devices are intended to measure the level of uniaxial compressive stress in the region of measurement. Little success has been achieved with the development of shear-stress-measuring sensors, although large devices have been reported for limited applications (Bennet *et al.* 1979; Tappin *et al.* 1980).

A number of electronic pressure transducers are available commercially. Most of these devices employ a strain-gauge diaphragm as their sensing surface and are designed to measure hydrostatic pressure in fluids; they should be used with caution when measuring uniaxial stress at the interface between compliant solids.

In order to reduce measurement errors, the ratio of thickness to diameter (aspect ratio) of suitable devices should be small and devices should ideally be capable of conforming to the curvature of the region being measured (Ferguson-Pell 1980). Most strain-gauged diaphragm devices do not fully meet these requirements.

A number of electro-pneumatic devices have been developed, the most widely used probably being the Talley-Scimedics Pressure Evaluator (Talley, Borehamwood, US; Scimedics, Anaheim, USA) developed by Reswick and Rogers (1976) which has an overall diameter of approximately 100 mm. Smaller sensors are also available (Talley, UK; TIRR, Houston, Texas, USA: Denne gauge, Sussex, UK).

In order to compare the performance of the electronic and pneumatic type transducers, a calibration apparatus was constructed. A dead-weight loading device consisting of an upper platen 200 mm in diameter, a 100 mm square sheet of PVC gel 15 mm thick to simulate flesh, an assortment of equal sized blocks of typical wheelchair cushion materials 25 mm thick, and a lower loading surface. Loads were applied to the upper platen, and the sensor to be tested was sandwiched between the upper PVC gel and the simulated wheelchair cushion material.

Two electronic sensors were tested, the Precision Model 156 (Precision Measurement Co., Ann Arbor, USA) and the Kulite LQS-125-200 (Kulite Semiconductor Products Ridgefield, NJ, USA). Comparison was made with similar tests on the Talley-Scimedics Evaluator and a small 25 mm diameter pneumatic sensor produced by TIRR.

Table 17.1 summarizes the results obtained, which sought to compare the response of the sensors with each other, with the known applied nominal stress, and for two different loading interfaces, gel–gel and gel–foam. The results indicate large divergences from the nominal applied stress, the pressures indicated by all the sensors appearing to read high. Marked differences were also observed between the two types of loading surfaces indicating consistently higher readings in the gel–gel configuration.

An underlying assumption of many studies that have used the Talley-Scimedics Evaluator has been that it measures peak pressure over the bony prominence being monitored (Reswick and Rogers 1976). A study undertaken by our group indicates that the device has some averaging effect in producing a reading.

Table 17.1. *Comparison of electronic and air cell transducers on surfaces of differing compliances and surface properties. The values indicated are means for five readings on each surface in kPa. The figures in parentheses are the corresponding standard deviation*

Nominal applied stress kPa [mm Hg]	Pressure readings (kPa)							
	Precision		Kulite		Talley-Scimedics		TIRR air cell	
	Foam	Gel	Foam	Gel	Foam	Gel	Foam	Gel
3.3 [26]	3.49 (0.24)	6.30 (0.25)	3.59 (0.10)	6.63 (0.46)	3.45 (0.11)	5.37 (0.19)	2.11 (0.29)	6.16 (0.41)
6.7 [52]	8.15 (0.33)	11.79 (0.34)	7.96 (0.23)	11.93 (0.82)	7.34 (0.08)	10.82 (0.20)	6.26 (0.56)	12.39 (0.40)
10.0 [78]	13.39 (0.69)	17.17 (0.46)	12.60 (0.33)	17.02 (0.32)	11.59 (0.14)	16.28 (0.21)	10.95 (0.64)	18.12 (0.34)
13.3 [104]	18.37 (0.79)	22.04 (0.53)	17.69 (0.47)	21.96 (0.25)	16.08 (0.16)	21.84 (0.11)	15.87 (1.04)	23.68 (0.47)
16.6 [130]	23.72 (1.13)	26.23 (0.81)	23.05 (0.58)	26.68 (0.33)	20.50 (0.11)	27.19 (0.13)	21.34 (1.48)	29.14 (0.25)
19.9 [156]	28.95 (1.29)	30.52 (0.83)	28.48 (0.69)	31.08 (0.44)	25.10 (0.11)	32.76 (0.11)	26.70 (1.80)	34.32 (0.40)

A comparison of sitting pressures on wheelchair cushions was made using a Talley-Scimedics Evaluator and a matrix of five Kulite strain-gauge diaphragm sensors. Three measurements were made on one able-bodied male volunteer using a series of 21 commercially available cushions. The Kulite sensors were positioned as indicated in Fig. 17.1, and the Talley-Scimedics sensor was placed under the ischium of the opposite buttock. The average of three repeated tests for sensors in the Kulite matrix was calculated and compared with the corresponding average of three readings from the Talley-Scimedics Evaluator. The Kulite sensors indicated a wide variation in pressure according to their position relative to the ischial tuberosity. The peak pressure measured with the Kulite sensor matrix was generally higher than the pressure indicated by the Talley-Scimedics device. Statistical comparison using the *t*-test between the mean pressure measured by the Talley-Scimedics sensor and the highest mean pressure reading from the Kulite matrix indicated a significant difference in the means for the three representative foams tested (Table 17.2).

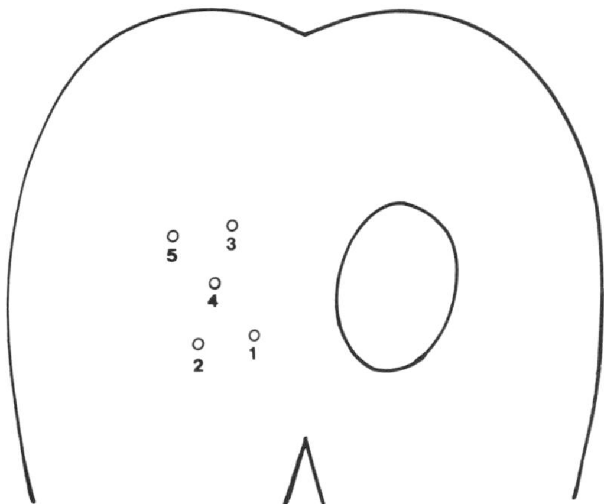

Fig. 17.1. Diagrammatic representation indicating relative positions of Kulite sensor matrix (left) and Talley-Scimedics Evaluator (right).

The average of the mean pressures from the five Kulite sensors was computed and compared with the mean of the Talley-Scimedics readings. Data analysed by this procedure indicate a close correspondence between the average pressure beneath the ischial tuberosity using the matrix and the value obtained with the Talley-Scimedics Evaluator (Palmieri *et al.* 1980).

A major difficulty in interpreting these results is due to the fact that neither sensor can be used as a reference against which the other can be tested. The results do however suggest that the Talley-Scimedics results should be interpreted to be more closely a measure of average pressure rather than peak pressure. The small size and high cost of electronic sensors prohibits their use for

Table 17.2. *Comparison of pressure readings obtained using a matrix of Kulite electronic transducers and a Talley-Scimedics Evaluator placed under left and right ischial tuberosities respectively of three able-bodied volunteers. The mean values of the Kulite transducers placed directly under the ischial tuberosity were compared with those obtained with the Talley-Scimedics Evaluator. They differ significantly at the* *p = *0.01 level;* **p = *0.01 level,* (t-test)*. The mean Kulite matrix pressure does not differ significantly from the mean Talley-Scimedics value. The figures in parentheses are the corresponding standard deviation*

Cushion type	Pressure readings (kPa)						
	Mean Kulite #1	Mean Kulite #2	Mean Kulite #3	Mean Kulite #4	Mean Kulite #5	Mean Kulite Matrix	Mean Talley-Scimedics
Foam	12.7 (1.3)	7.4 (2.8)	12.0 (1.6)	13.8* (3.2)	10.7 (0.9)	11.3 (0.6)	11.9 (0.5)
Visco-elastic Foam	9.2 (1.0)	7.6 (2.4)	10.2 (1.4)	11.2** (1.4)	9.4 (1.2)	9.5 (0.7)	9.0 (0.2)
Gel	12.4 (1.6)	10.3 (2.8)	12.2 (1.6)	14.0** (1.0)	13.2 (1.7)	12.4 (0.7)	12.3 (0.8)

routine wheelchair cushion fitting in the clinic. The average effect of the Talley-Scimedics device suggests that considerable caution should be exercised in interpreting its output over concentrated pressure distributions. Devices helpful in using and positioning multiple sensors for clinical measurements have also been developed at Helen Hayes Hospital (Mayo-Smith 1980; Mayo-Smith and Cochran 1981).

In addition to measuring interface pressure, a technique developed by Snashall *et al.* (1971) was adopted to measure interstitial pressure approximately 2–5 mm beneath the surface of the skin using wick catheters and compared with the value of interface pressure using the Talley-Scimedics Evaluator.

Results were obtained by inserting the wick catheters into the posterior thighs of three human volunteers and were then compared with sequential measurements of interface pressure using the Talley-Scimedics Evaluator. The subject was seated on a medium density foam (Rogers 3040, USA) with feet hanging free. In order to observe the correspondence between interface and interstitial pressures over a range of pressures, weights of 6.8 kg and 11.4 kg were placed on the feet. The mean interstitial pressure for the three subjects obtained with the wick catheter transducer and the interface pressure measured by the Talley-Scimedics Evalutor were compared using the t-test for each loading configuration and found to be in close agreement. The measurements were made beneath the thighs where there is a relatively uniform distribution of pressure. The results (Reddy *et al.* 1984) indicate that under these circumstances, a correspondence exists between interstitial pressure and interface pressures.

In laboratory experiments, the reliability of the wick method was studied in an animal model using pigs (Reddy *et al.* 1981). It was found that approximately 70 per cent of the externally applied pressure is transmitted to the transducer with a circumferential pressure cuff in normally hydrated states. However, if the microenvironment of the wick is altered by injecting small amounts of fluid (0.02 ml), then nearly 100 per cent of the externally applied pressure is transmitted to the fluid. When fluid is injected, it probably creates a pocket around the wick so that the total normal tissue stress is transmitted to the wick. In this state, the fluid pressure in the wick equals the total normal stress in the tissue rather than the tissue hydrostatic pressure. In the present experiments, fluid was injected into the tissue to ensure the measurement of the total normal stress. Care was taken to place the catheter close beneath the surface of the skin to minimize effects of stress distribution by the tissue.

The loads placed on the feet were selected to generate pressures underneath the thighs in the same range as those observed in the buttocks during sitting. The fact that the wick pressure (subcutaneous tissue pressure) correlated well with the normal stress at the interface as measured by the Scimedic's air-cell transducer supports its use as a measurement tool, although it obviously indicates average surface rather than peak pressures within its area. Although interface and subcutaneous measurements in the thighs correlated well, it would be difficult to accomplish the same experiment in the buttocks where the local position of the wick catheter in relation to the ischium would be difficult to determine.

MEASUREMENT OF THERMAL PARAMETERS

The measurement of skin temperature is easily achieved using modern thermocouples and thermistors. The mapping of skin temperature is, however, more complex and best achieved with infrared thermography or flexible sheets of material coated with temperature-sensitive liquid crystals.

A particular interest of the Helen Hayes Soft Tissue Mechanics Group has been the evaluation of the thermal dissipating properties of commonly used wheelchair cushion materials (Stewart *et al.* 1980). Simple discrete temperature readings were considered to be appropriate in this study and 'Hy Cal' TC-2345 (Santa Fe, CA, USA) thermocouples were used in conjunction with a Honeywell Electronic 16 (Westfield, NJ, USA) signal conditioning unit.

Skin temperatures were measured under the left and right ischial tuberosities of an able-bodied volunteer with thermocouples attached to the skin with a single strip of surgical tape. Room temperature was monitored continuously and maintained at 21.4 °C ± 0.1 °C.

Figure 17.2 represents the results from these experiments for five different cushion types. The foam cushions were not covered, and the water, gel, and

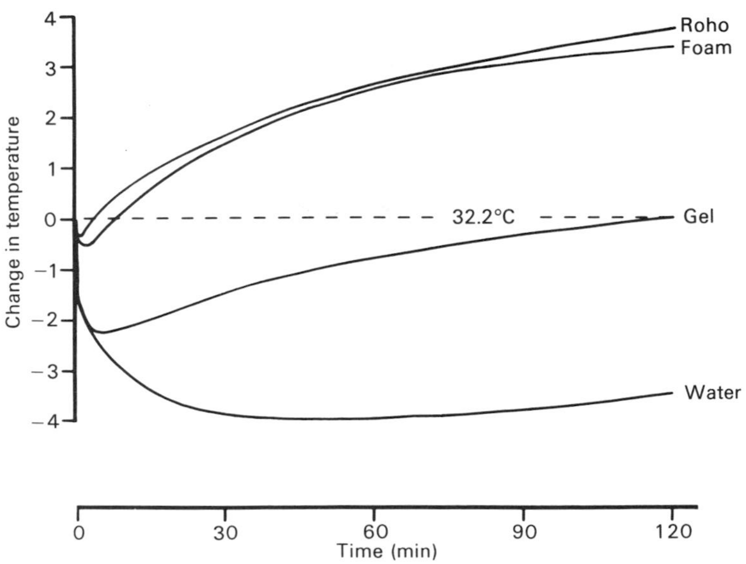

Fig. 17.2. Changes in skin temperature for a subject seated on cushions with differing thermal dissipation properties.

Roho cushions were tested without additional covering. The subject wore a pair of loose fitting cotton trousers.

The results suggest that foams and viscoelastic foams tend to be hot and increase skin temperature by several degrees, because the foam materials themselves and the air entrapped in them tend to be poor absorbers and conductors of heat. Gel pads produced a distinct initial cooling of the skin which had returned to close to the initial skin temperature after two hours. Water flotation pads provided a significant drop in skin temperature which was partially maintained throughout the two hours of sitting. The Roho cushion provided characteristics that were similar to those of the foam cushion.

In addition to skin temperature, the humidity at the interface between the skin and cushion can cause considerable discomfort and predispose a patient to skin damage caused by abrasion.

Close to the temperature sensor on the right buttock used in the above study, a relative humidity sensor was attached to the skin enclosed in a ventilated plastic housing to prevent contamination. The sensor (Phys-Chemical Research Corp., New York, USA) consisted of a small styrene block with an electrically conductive surface layer whose impedance is related to moisture absorption. The sensor was readable only above 50 per cent RH, which was an acceptable range for this study.

An alternative chemical sensor (Humidial Corp., Colton, CA, USA) was also employed close to the thermocouple on the left buttock. The sensor consists of a 25 mm diameter disc of blotting paper labelled with RH readings in increments of 10 from 20 to 90 RH. Each label is impregnated with a spot of cobaltous chloride solution that changes from blue to pink at the specified humidity. Because a lavender colour indicates an intermediate value of relative humidity, this sensor can be read to the nearest 5 per cent. For use beneath seated patients, the sensors were prepared according to Trandel and Lewis (1975) in a ventilated foam ring to prevent direct contact contamination. The readings from the two sensors corresponded closely at a 99 per cent level of confidence although the Humidial sensor was found in general to be easier to use. Table 17.3 indicates the range of changes in RH over a two-hour period of sitting for each of the cushions tested. The gel, water, and Roho cushions all generated higher changes in RH during a two-hour test than the foams despite the generally lower skin temperatures observed for the gel- and water-filled cushions. All the cushions employing impermeable surfaces in close contact with the skin were found to produce large increases in interface humidity, whereas the foams permit some air exchange owing to their porous construction, resulting in a smaller increase in RH despite being poorer thermal dissipaters.

SUMMARY

The evaluation of wheelchair cushions for patients vulnerable to pressure sores is complex and at present no one cushion appears to have all the characteristics

Table 17.3. *Changes in temperature and relative humidity for different cushion types over a two hour test period*

Cushion	Changes in skin temperature (°C)			Increase in % relative humidity
Minutes	30	60	120	120
Foam Rogers 1836	1.45	2.6	3.4	15
Viscoelastic Foam Alimed 164–S	1.2	2.3	3.25	17
Gel Resto Flo Pad 1565	−1.5	−0.85	0.1	39
Water Floatation Med Pro H_2O	−3.8	−.3.9	−3.4	36
Special Roho Balloon	1.65	2.65	3.75	34.5

required to suit a majority of users. Quantitative methods for comparing the properties of cushions and fitting them to individuals are essential, but the devices used for measurements on subjects sitting on cushions are difficult to calibrate and evaluate for scientific and clinical applications. Pressure sensors, in particular, may behave as discrete sensors of localized pressure or integrators of pressure distributions indicating average pressures. Some may do both depending upon the mechanical properties of the materials present at the interface.

In the studies reported, two types of pressure sensors were tested, the strain-gauged diaphragm and the air-cell type. Measurement of interstitial pressures were also made and compared with those at the interface. In bench calibrations, both the electronic and pneumatic sensors gave readings greater than the applied nominal stress and responded differently when inserted between materials of different compliances.

A comparison of a matrix of five Kulite electronic sensors and a Talley-Scimedics Evaluator placed beneath the ischial tuberosities of a volunteer indicated that the electronic sensors could identify a peak pressure well in excess of the value indicated by a carefully placed Talley-Scimedics Evaluator. The average of the Kulite matrix indicated a close correspondence with the Evaluator reading.

Interstitial pressure measured with a wick catheter placed 2–5 mm beneath the skin's surface under the thigh of a group of three volunteers were found to correspond closely with those measured at the interface using the Talley-Scimedics Evaluator.

Temperature and humidity measurements beneath a seated subject were monitored to study the termal characteristics of different cushions. Gels and water-filled cushions appeared to reduce skin temperature significantly compared with foam and the Roho cushions. Relative humidity was also measured and found to be considerably higher on cushions covered with impermeable covering materials.

It is clear even from these limited studies that no one cushion is ideal from all standpoints. For difficult problems, the proper cushion or cushion material combinations must be prescribed for the individual patient. In this process, not only mechanical properties but thermal as well should be considered, and in many instances, compromise will be necessary with associated compensatory action by the patient through limiting sitting times or increasing pressure relief frequency.

Acknowledgements

This work was partially supported by: National Institute for Handicapped Research through Grant NIHR (RSA) 23-P-59173; Veterans Administration Contract V5244-P-1375; Rehabilitation Services Administration, Department of HEW, Washington, DC; Texas Institute for Rehabilitation and Research, NIHR Grant No. 23-P-57888.

REFERENCES

Bailey, B. N. (1967). *Bedsores.* Edward Arnold, London.

Bennett, L., Kavier, D., Bok, K. L., and Travier, F. A. (1979). Shear vs. pressure as causative factors in skin blood flow occlusion. *Archs phys. med. Rehabil.* **60**, 309–14.

Berecek, K. H. (1975). Etiology of decubitus ulcers. *Nurs. Clins N. Am.* **10**(9), 157–69.

Brown, A. C. and Brengelmann, G. (1965). Energy metabolism. In *Physiology and biophysics* (ed. R. C. Ruch and H. D. Pattern) pp. 1030–79. Saunders, Philadelphia.

Cochran, G. V. B. and Palmieri, V. (1980). Development of test methods for evaluation of wheelchair cushions. *Bull. prosthet. Res.* **17**, 9–30.

Daniel, R. K., Priest, D. L., and Wheatley, D. C. (1981). Etiologic factors in pressure sores: an experimental model. *Archs phys. med. Rehabil.* **62**, 492–8.

Ferguson-Pell, M. W. (1980). Design criteria for the measurement of pressure at body/support interfaces. *Engng Med.* **9**, No. 4.

Fernie, G. R. (1973). Biomechanical aspects of the aetiology of decubitus ulcers on human patients. Ph.D. thesis, University of Strathclyde, Glasgow.

Groth, K. E. (1942). Klinische Beobachtungen und Experimentelle Studen Uber Die Enstehung Des Dekubitus. *Acta chir. scand.* **87**, Suppl. 76.

Guttmann, L. (1976). The prevention and treatment of pressure sores. In *Bedsore biomechanics* (ed. R. M. Kenedi, J. Cowden, and J. T. Scales). Macmillan, London.

Hussain, T. (1953). An experimental study of some pressure effects on tissues, with reference to the bed-sore problem. *J. path. Bact.* **66**, 347–58.

Kosiak, M. (1959). Etiology and pathology of ischemic ulcers. *Archs phys. med. Rehabil.* **40**, 62–9.

— (1961). Etiology of decubitus ulcers. *Archs phys. med. Rehabil.* **42**, 19–29.

Mayo-Smith, W. (1980). Pressure measurements: device for obtaining readouts from multiple sites using pneumatic transducers. *Archs phys. med. Rehabil.* **61**, 460–1.

— and Cochran, G. V. B. (1981). Wheelchair cushion modification: device for locating high-pressure regions. *Arch phys. med. Rehabil.* **62**, 135–6.

Palmieri, V. R., Haelen, G., and Cochran, G. V. B. (1980). A comparison of sitting pressures on wheelchair cushions as measured by 'air cell' transducers and miniature electronic pressure transducers. *Bull. prosthet. Res.* **17**, 5–8.

Reddy, N. P., Palmieri, V., and Cochran, G. V. B. (1981). Subcutaneous interstitial fluid pressure during external loading. *Am. J. Physiol.* **240** (Regulatory Integrative Comp. Physiol. 9), R327–9.

— — — (1984). Evaluation of transducer performance for buttock–cushion interface pressure measurements. *J. Rehab. Res. Devel.* **21**, 43–50.

Reswick, J. B. and Rogers, J. E. (1976). Exprience at Rancho Los Amigos Hospital with devices and techniques to prevent pressure sores. In *Bedsore biomechanics* (ed. R. M. Kenedi, J. M. Cowden, and J. T. Scales) pp. 310–10. Macmillan, London.

Schell, V. C. and Wolcott, L. E. (1976). Etiology, prevention and management of decubitus ulcers. *Mo. Med.* **63**, 109–12.

Snashall, P. D., Lucas, J., Guz, A., and Floyer, M. A. (1971). Measurement of interstitial fluid pressure by means of a cotton wick in man and animals: an analysis of the origin of pressure. *Clin. Sci.* **41**, 35–53.

Stewart, S. F. C., Palmieri, V. R., and Cochran, G. V. B. (1980). Wheelchair cushion effect on skin temperature, heat flux, and relative humidity. *Archs phys. med. Rehabil.* **61**, 229–33.

Tappin, J. W., Polland, J., and Beckett, E. A. (1980). Method of measuring shearing forces on the sole of the foot. *Clin. Phys. physiol. Movement* **1**.

Torrance, C. (1981). Pressure sores: pathogenesis, prophylaxis and treatment. *Nurs. Times* **77**, 5–8.

Trandel, R. S. and Lewis, D. W. (1975). Small pliable humidity sensor with special reference to prevention of decubitus ulcers. *J. Am. geriat. Soc.* **23**, 322–6.

18 Pressure Measurement at the Patient–Support Interface

D. L. BADER, J. GWILLIM, T. P. NEWSON, AND DEREK HARRIS

INTRODUCTION

The use of an orthotic device, to provide functional support to the musculoskeletal system of the body, will necessarily involve the transmission of forces through the soft tissues. These forces, applied across specific contact areas, may be sufficient to produce substantial stresses within the soft tissues, which can impair the blood supply and lymphatic drainage. If these interface conditions are prolonged, cell necrosis will result and may lead to the eventual development of tissue breakdown and ulceration. Common areas of risk are at the seating–mattress interface and under the plantar surface of the foot. Patients particularly at risk are those with restricted mobility of both a temporary and a permanent nature. The latter group involves patients with chronic neuromuscular disease, such as multiple sclerosis, or traumatic injuries of the spinal cord. The former group includes those having operations, where soft tissues may be exposed to pressure for an extended period during anaesthesia. For example, elderly patients who are undergoing repair of fractures of the femur often develop pressure sores following surgery. Thus the monitoring of pressure, and more especially of pressure distribution, at the patient support interface is of paramount clinical importance in assessing the viability of soft tissues. This chapter discusses the development of such a system.

EXISTING TECHNIQUES

In the design of an interface pressure transducer, certain specifications should be fulfilled (Bader 1982). Foremost, its presence must not affect the parameter being measured. This implies that the transducer in contact with the soft tissues should be more compliant than either the soft tissues or the support orthosis. In addition, such a device should be robust, but flexible enough to be used on curved body contours.

In the literature there are reports of various pressure measurement devices employing, for example, electropneumatic, pneumatic, capacitive, and resistive techniques. These have been reviewed by Ferguson-Pell *et al.* (1976). However, no single system meets all the specifications demanded of an interface transducer and achieves suitable performance in terms of sensitivity, linearity, negligible hysteresis, and temperature independence.

The Denne gauge (Huntleigh Medical Ltd) is one of the few commercial clinical pressure sensors. It is a simple pneumatic device composed of an inflated compliant bag connected to a mercury column. It has the obvious advantages of being thin and flexible, and it is unaffected by the chemical environment encountered in the clinical situation. A pneumatic system is also inherently insensitive to shear forces and temperature changes, which can be present at the patient interface.

PRINCIPLE OF SYSTEM

The pneumatic principle forms a basis for the present development. Initially, a system similar to the Denne gauge (as described by Crewe 1983) was modified to allow a deflated bag to be introduced at the patient interface. A volume of air was then injected into the bag via a three-way tap, and the pressure measurement obtained. This finite volume of air determines not only the thickness of the bag, but also its accuracy to measure a specific applied pressure. The inevitable errors which arise may be estimated from the family of curves in which pressure is monitored while air is introduced into the system at a constant rate. The five curves in Fig. 18.1 represent five conditions of loading the bag. The form of these curves has been explained by O'Leary and Lyddy (1978), who divided them into three distinct phases. Initially the pressure in the system increases linearly up to the applied pressure (phase A). During this phase the bag remains flat at the interface and the incoming air only pressurizes the connecting air lines. As the applied pressure is reached and exceeded, the bag starts to inflate, the system volume increases and the rise in pressure is smaller for a given injected volume of air (phase B). The exact slope of the curve is largely determined by the compliance of the interface materials. In the final phase the slope of the curves increases dramatically, as the bag has become fully inflated and the resulting tension in its walls resists further changes in volume.

DESCRIPTION OF SYSTEM

This characteristic behaviour has been employed in the present system. It uses the minimum injected volume of air to measure an applied pressure, and does not appreciably deform the interface under investigation. This has been achieved by pressurizing the system with a constant mass flow rate, as controlled by high-pressure pump and a needle value. The pressure/time characteristics of the system are continuously monitored with an electronic circuit. This is programmed to detect the point of changeover from phase A to phase B, when the system pressure equals the applied pressure and the bag starts to open. At this instant of time, a static measurement of system pressure is obtained using a

A		Applied pressure is greater than pneumatic pressure
B		Applied pressure is equal to pneumatic pressure
C		Applied pressure is less than pneumatic pressure

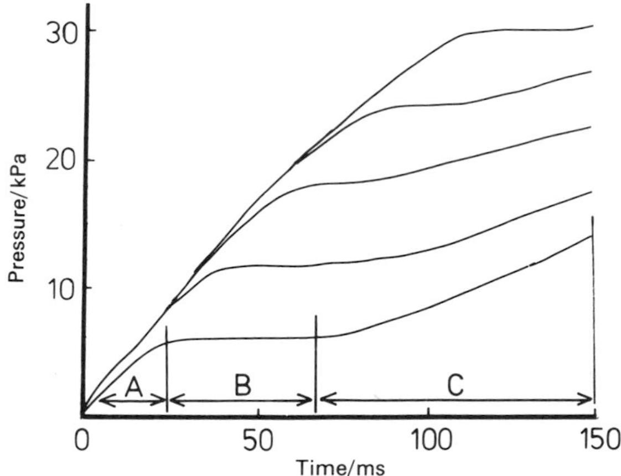

Fig. 18.1. The pressure–time characteristics of an individual pneumatic cell, subjected to five different applied pressures. A, B, and C refer to the three phases of inflation (see text).

semiconductor strain-gauge pressure transducer (range 0–100 kPa; 0–750 mm Hg). The air flow is then reversed and the system is exhausted. The frequency of the total cycle necessarily depends upon the applied pressure, but is certainly in excess of 2 Hz at pressure of 33 kPa, the present upper limit of the instrument.

In conjunction with the development of this instrument, the use of a single bag or cell has been replaced by a matrix of cells. The matrix has the advantage of indicating both high-pressure areas and the pressure distribution at the interface. This assessment is important as the presence of pressure gradients has been implicated as potentially detrimental to the viability of soft tissues (Husain 1953). The technique employed to measure the interface pressure is ideally suited for matrix application, as all cells not under investigation remain deflated.

This prevents the distortion of the pressure distribution which occurs when the subject locally floats on a cushion of air cells. A manual switch is incorporated in the instrument to permit investigation of the individual cells in the matrix.

The design of the matrix has followed from both theoretical considerations and practical testing of conditions at the patient interface. It takes the form of a four-by-three rectangular matrix of 12 cells. The individual cells, 20 mm in diameter, are spaced at a distance of 28 mm between centres. This matrix, covering a total area of 8000 mm^2, may be used at various seating areas of interest, as well as other interfaces with orthotic and prosthetic devices.

The present system monitors the interface pressure in the 12 individual cells. The pressure in these cells is stored in the microprocessor-controlled instrument and displayed in a numeric map of the matrix. There is also an analogue scale to visualize individual pressure readings. The complete system is illustrated in block diagram form in Fig. 18.2.

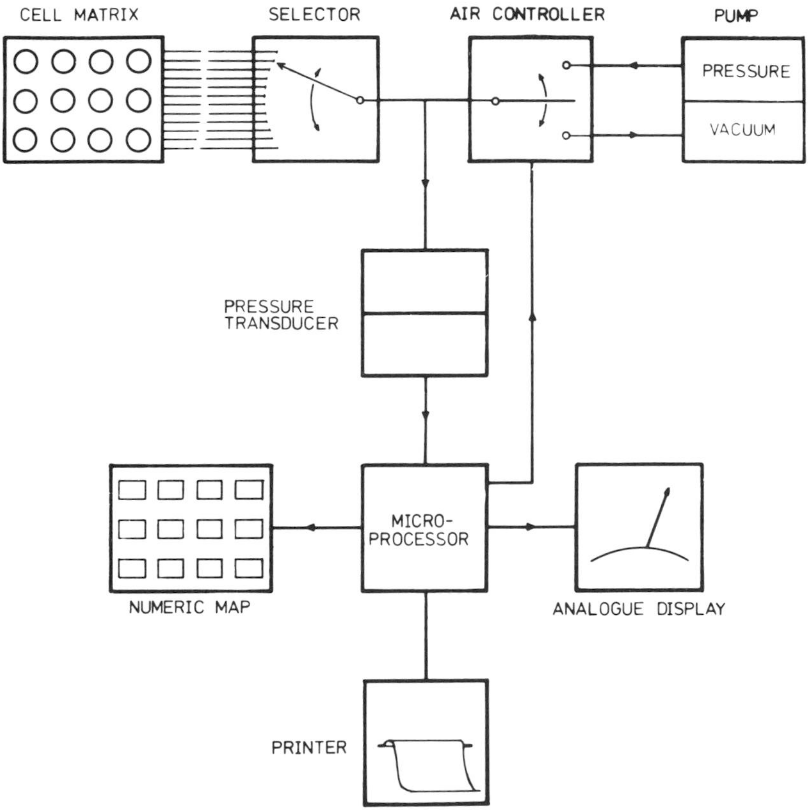

Fig. 18.2. A block diagram of the complete system.

CALIBRATION

Two methods were employed to calibrate the instrument, following the suggestions of Ferguson-Pell (1980). The first involved the use of a sealed pressure chamber and an extruded polyurethane membrane, which permits the transmission of uniaxial forces to a pneumatic cell. Results indicated a linear response of the cell up to a pressure of about 33 kPa (Fig. 18.3).

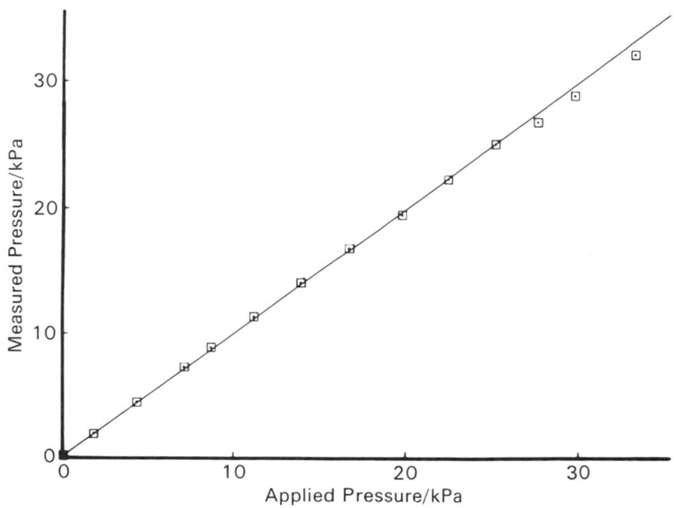

Fig. 18.3. The response of a pneumatic cell to an applied uniaxial pressure.

The second method closely simulates the loading conditions encountered at the patient interface. The flexible matrix of 12 cells was sandwiched between two layers of viscoelastic foam, to simulate the soft tissues and the support material, in an Instron 1122 testing machine. Loading through a large flat indenter permitted a comparison of the behaviour of the individual cells in the matrix. A maximum difference of ±6 per cent was observed at any nominal pressure up to 33 kPa. In addition the matrix was subjected to uniaxial loading using an indenter of diameter 180 mm. Results indicated that the design of the matrix was sensitive to pressure gradients up to a value of 0.45 kPa/mm. This is comparable to maximum values estimated at the seating interface (Ferguson-Pell 1980).

CONCLUSIONS

The developed system provides an accurate measurement of pressure distribution at the patient support interface. It employs a pneumatic principle to monitor the pressure continuously in any of a matrix of 12 cells. This instrument is intended for clinical use in the assessment of tissue viability.

REFERENCES

Bader, D. L. (1982). Interface pressure measurement. *Care Sci. Pract.* **1**, 22–4.

Crewe, R. (1983). The role of the occupational therapist in pressure sore prevention. In *Pressure sores* (ed. J. C. Barbanel, C. D. Forbes, and G. D. Lowe) pp. 121–32. Macmillan, London.

Ferguson-Pell, M. W. (1980). Design criteria for the measurement at body/support interfaces. *Engng Med.* **9**, 209–14.

—, Bell, F., and Evans, J. H. (1976). Interface pressure sensors. In *Bed sore biomechanics* (ed. R. M. Kenedi) pp. 189–97. Macmillan, London.

Husain, T. (1953). An experimental study of some pressure effects on tissue with reference to the bed sore problem. *J. path. Bact.* **66**, 347–58.

O'Leary, J. P. and Lyddy, T. E. (1978). A non-distorting transducer for measuring pressure under load-bearing tissue. In *Proceedings of the 5th Annual Conference of Systems and Devices for the Disabled*, pp. 239–42. Baylor College of Medicine, Houston, Texas.

19 Skin Viability Measurement Using the Transcutaneous Oxygen Monitor

G. S. E. DOWD, K. LINGE, AND G. BENTLEY

INTRODUCTION

A common unresolved problem for surgeons is determination by clinical examination whether skin healing will occur following operation in patients with known or suspected vascular deficiency in the lower limbs. Necrosis of skin following operation will result at best in prolonged recovery and at worst in amputation of the limb. In the elderly, foot deformities often require a surgical procedure for correction. These patients frequently have peripheral vascular deficiency, yet careful clinical examination of the skin circulation including palpation of peripheral pulses, assessment of capillary return and colour of the skin may not allow the surgeon to decide whether the circulation would withstand the trauma of an operative procedure. As a result, many patients have probably been refused operation and others have undergone operation in which the skin has failed to heal. In patients with peripheral vascular disease severe enough to require amputation, the site of amputation depends upon the state of skin viability and on the site of optimal limb function following amputation. Clinical examination may provide some information on the site in regard to primary stump healing, but often the surgeon is presented with a patient who has a cold, pale leg in which it is impossible to decide whether a more distal amputation will heal. The cautious surgeon will perform a more proximal amputation associated with a greater chance of stump healing, but with a poorer functional potential (Waters *et al.* 1976).

ASSESSMENT OF SKIN VIABILITY

Skin viability depends on many factors, but the ability of the circulation to provide an adequate nutrition is paramount. Many methods have been developed to provide a precise measurement of the circulation in the cutaneous tissues. However, many of the techniques used produce measurements not directly related to the area of skin under examination. For example, Doppler segmental blood pressure measurements in the lower limbs are directly related to the main arteries, including the popliteal, posterior tibial or dorsalis pedis, whereas the measurement is used clinically to predict healing in the skin some distance away (Yao *et al.* 1969; Dean *et al.* 1975). The inference is that the haemodynamic state of main arteries directly affects local cutaneous tissue. In many cases this is true, but in others, the inference is not valid. A further requirement of the Doppler segmental blood pressure technique is the presence of peripheral pulses, yet it has been shown that skin will heal distal to a main vessel in which a pulse cannot be palpated (Sarmiento *et al.* 1970; Mooney *et al.* 1976).

Direct measurement of the skin vessel perfusion pressures should obviate these problems and more recent methods of assessing skin viability have been directed to measurement of this factor. Techniques to measure flow have included fluorescein angiography (Horne and Tanser 1982), laser Doppler (Holloway 1980), and measurement of skin temperature using thermography (Robinson 1981). Macro-aggregated albumin microspheres labelled with iodine-131 have been injected intra-arterially and their distribution in the cutaneous tissues measured with a Gamma camera using a pulse-height analyser (Siegel *et al.* 1975). The latter technique has been performed on a small number of patients with peripheral vascular disease, but the results are inconclusive. The technique, which is similar to lung perfusion scanning, is potentially useful and warrants further investigation. Methods for investigating skin blood perfusion pressure have also been developed using radioactive isotopes injected intradermally, as a depot, and the applied pressure at which the radioisotope fails to be cleared from the site by the circulation is measured with a sphygmomanometer cuff (Holstein *et al.* 1978). The method has been shown to be a useful way of measuring the pressure above which skin healing is likely to occur and has been used in the clinical situation in deciding on above or below knee amputations for peripheral vascular disease. Problems associated with the method include the fact that the technique is invasive and requires some degree of expertise to perform. It may cause the patient discomfort and pain due to compression of the limb by the sphygmomanometer. The photoelectric cell has been used as an alternative method of measuring skin blood pressure by measuring the applied pressure at which the blood cells cease to flow when pressure is applied to the skin (Holstein *et al.* 1979). A similar method has been used, using the laser Doppler machine to demonstrate cessation of flow against applied pressure (Holloway 1980).

The problems and limitations, associated with many of the techniques described above, are that the results may not be accurate or the technique may be difficult to apply to the clinical situation and may damage the cutaneous tissues which already are potentially at risk. Other methods are not quantitative. As a result of our dissatisfaction with the methods available for determining skin viability, the ability of the transcutaneous oxygen monitor to act as a means of assessing skin viability has been investigated.

TRANSCUTANEOUS OXYGEN MONITORING

Gerlach (1851) was the first to show that an exchange of oxygen and carbon dioxide occurred between the skin and the air. In 1951, Baumberger and Goodfriend, working at Stanford University, showed that the oxygen diffusing through the skin of the finger placed in a warm phosphate buffer solution at 45 °C was equal to the arterial oxygen tension. Clark (1956) described an electrode in which neither the cathode nor the anode came into contact with the solution in which the oxygen was dissolved. By using a modified Clark electrode, Huch *et al.* (1969) showed that it was possible to obtain direct readings of the transcutaneous oxygen tension (TC PO_2), by a polarographic method, and that by heating the skin to the point of maximal vasodilatation, the TC PO_2 could be used as a practical method of continuous arterial oxygen pressure monitoring (PAO_2). This technique was originally used to monitor the arterial oxygen tension in neonates, but has since been used in adult intensive care units. Several subsequent papers observed that the TC PO_2 was related to the arterial perfusion pressure. Peabody *et al.* (1978) and Rooth *et al.* (1957) have shown that when the blood pressure falls below 90 mm Hg the TC PO_2 and PAO_2 become dissociated. Further investigators have shown that there is a relationship between the TC PO_2 and perfusion pressure in the cutaneous tissue (Wyss *et al.* 1981; Dowd 1982). As the perfusion pressure in the cutaneous tissues falls the TC PO_2 decreases (Fig. 19.1). The relationship between the TC PO_2 and perfusion pressure is such that it is a sensitive index of low peripheral pressure in the cutaneous tissues. Since skin ischaemia presents clinically when the perfusion pressure falls below 50-60 mm Hg, then the transcutaneous oxygen monitor has the potential to provide accurate measurement of this factor (Dean *et al.* 1975).

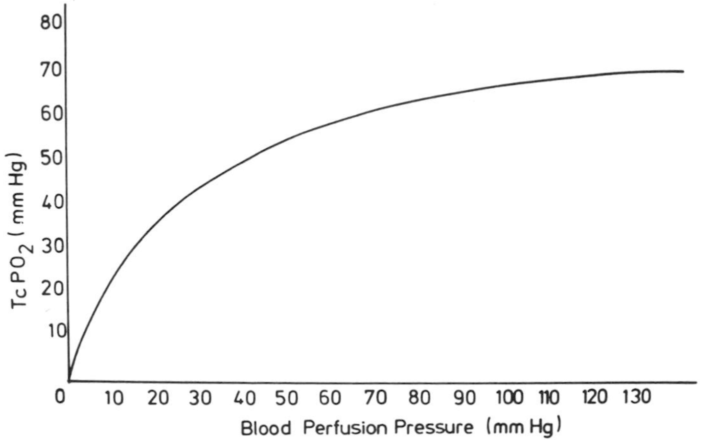

Fig. 19.1. Relationship between blood perfusion pressure and TC PO_2.

MEASUREMENTS IN NORMAL VOLUNTEERS

In order to evaluate the transcutaneous oxygen monitor in states of skin ischaemia, baseline data were obtained from the lower limbs of 161 normal volunteers. In each volunteer, a Radiometer TCMI TC oxygen monitor was used, with the heater of the electrode set at 44 °C to obtain maximal vasodilatation without burning the skin. Measurements were made in a relatively draught-free room. The volunteers were not allowed to talk during measurement. Each subject was rested supine on a couch for 20 minutes before measurement. The electrode was calibrated to 0 mm Hg and atmospheric oxygen tension, then attached to the skin at fixed points on the lower limb: dorsum of the foot, 10 cm below the knee joint, lateral to the anterior tibial border, and 10 cm above the knee (Fig. 19.2). The TC PO_2 was taken at the point at which the reading stabilized, usually 20 minutes after application of the electrode.

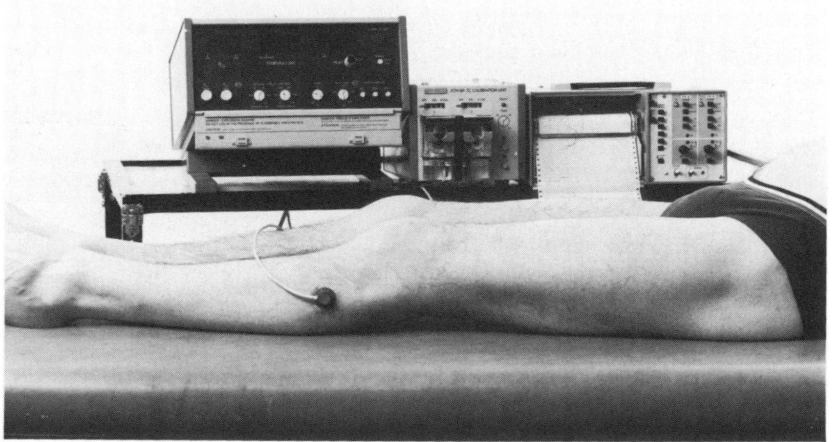

Fig. 19.2. Transcutaneous oxygen measurement in normal volunteer.

The normal range of TC PO_2 was 45–95 mm Hg with a mean of 70 mm Hg in 161 volunteers (Dowd *et al.* 1983*a*). The TC PO_2 measured in the lower limb was almost independent of the age and sex of the volunteer (Dowd *et al.* 1983*b*). There was a minimal TC PO_2 gradient from proximal to distal part of the limb in each individual (Ohgi *et al.* 1981; Dowd 1982).

MEASUREMENTS IN PATIENTS WITH PERIPHERAL VASCULAR DISEASE

In order to correlate TC PO_2 measurements with varying clinical degrees of skin ischaemia, a group of patients with ischaemic changes in the lower limbs were examined using the monitor. The patients were placed into three groups: those with symptoms of intermittent claudication, but without obvious skin changes in the lower limb; those with ischaemic changes in the limb periphery but without gangrenous changes, and those with frank gangrene of the skin of the foot. Measurements were taken on the dorsum of the foot and it was shown that the range of TC PO_2 measurements decreased with increasing severity of the ischaemia (Fig. 19.3).

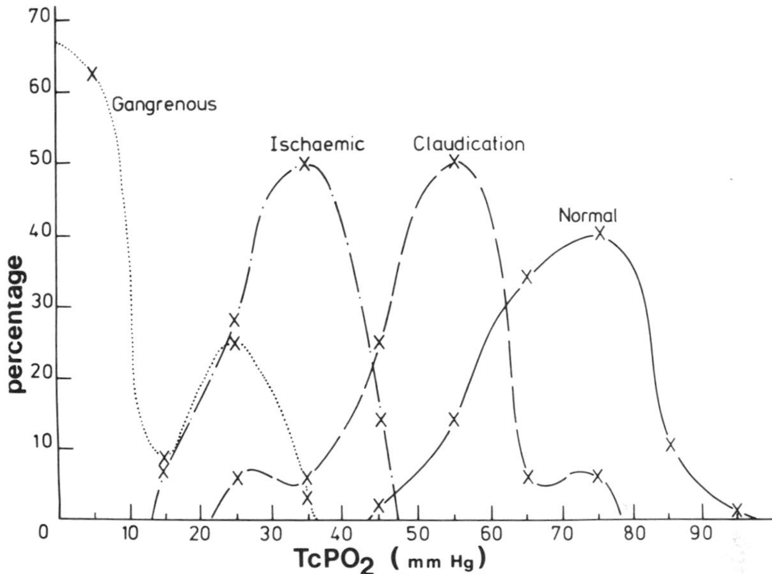

Fig. 19.3. Distribution of TC PO_2 on the dorsum of the foot in normals and three skin groups (see text).

MEASUREMENTS PRIOR TO AMPUTATION

The transcutaneous oxygen monitor was then used to measure the TC PO_2 at fixed sites in the lower limb of patients about to undergo amputation for peripheral vascular disease. The sites measured were the classical sites for above knee amputation, below knee amputation, and midtarsal amputation. The results of the measurements were not revealed to the operating surgeon who decided on the site of amputation on clinical grounds. A series of 51 patients were examined prior to operation and it was found that if the TC PO_2 level at the site of amputation was above 40 mm Hg then the skin flaps would heal (Fig. 19.4). The site of amputation did not affect this figure and was the same for above knee, below knee, and midtarsal amputations.

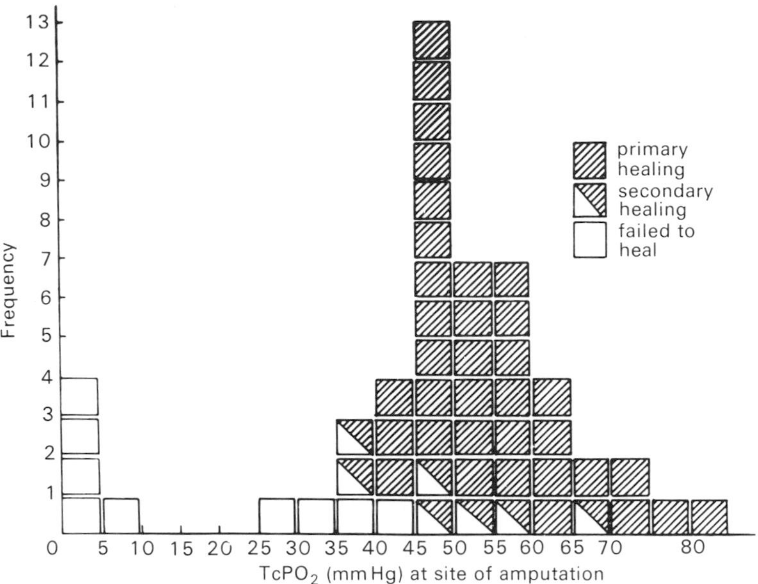

Fig. 19.4. Relationship between TC PO_2 at amputation site and skin-flap healing.

A comparison was made between the predicted successful site of lower limb amputation based on the transcutaneous oxygen levels and the site of amputation based on clinical assessment. Twenty-three patients who had an above-knee amputation performed healed by primary intention. Review of the preoperative TC PO_2 levels at the below knee site showed a TC PO_2 level greater than 40 mm Hg in 13 patients and that in all probability a below-knee amputation could have been successfully performed with an increased functional result to the patient. Of the 10 patients in which a below-knee amputation had been performed, comparison between the preoperative TC PO_2 level at the dorsum of the foot and below-knee levels showed that in all cases the midtarsal readings were below the level required to obtain skin healing. In other words none of the patients who had a below-knee amputation could have had a successful midtarsal amputation based either on clinical assessment or TC PO_2 levels.

The transcutaneous oxygen monitor does appear to be an accurate method of assessing the state of skin viability and the potential of skin flaps to heal. These results have been substantiated by Burgess *et al.* (1982) who have shown that a TC PO_2 greater than 40 mm Hg will result in healing of below-knee amputation stumps. The measurement of transcutaneous oxygen tension has been performed in patients undergoing Keller's operations on the great toe for hallux valgus and all those patients examined with a TC PO_2 greater than 40 mm Hg on the dorsum of the foot have healed. Only one patient, with a TC PO_2 of 37 mm Hg, has failed to heal and unfortunately required a below-knee amputation.

The use of the transcutaneous oxygen monitor in cases of injury is being investigated and has been shown to be useful in the examination of damaged skin associated with fractures of the tibia. At present, however, no definitive statement can be made concerning such cases. Both Achauer *et al.* (1980) and Keller *et al.* (1978) have shown that the TC PO_2 monitor is useful in investigating the viability of skin flaps in rabbit skin allografts.

CONCLUSIONS

The transcutaneous oxygen monitor has enabled decisions to be made which are more precise than those achieved by any other clinical or technical method. This method, which is non-invasive, has proved to be a reliable technique for assessing the state of skin viability in cases of chronic ischaemia. It allows direct measurement of small area of skin, resulting in a more confident approach to the assessment of skin viability than can be obtained by clinical examination, or by most other methods presently available.

REFERENCES

Achauer, B. M., Black, K. S., and Litke, D. K. (1980). Transcutaneous pO_2 in flaps: a new method of survival prediction. *J. plastic reconstruc. Surg.* **65**, 738–45.

Baumberger, J. P. and Goodfriend, R. B. (1951). Determination of arterial oxygen tension in man by equilibration through intact skin. *Fedn Proc. Fedn Am. Socs exp. Biol.* **10**, 10.

Burgess, E. M., Matsen, F. A., Wyss, C. R., and Simmons, C. W. (1982). Segmental transcutaneous measurements of pO_2 in patients requiring below-the-knee amputation for peripheral vascular insufficiency. *J. Bone Jt Surg.* **64A**, 378–82.

Clark, L. C. (1956). Monitor and control of blood and tissue oxygen. *Trans. Am. Soc. Art. Int. Org.* **2**, 41–8.

Dean, R. H., Yao, J. S. T., Thompson, R. G., and Bergan, J. J. (1975). Predictive value of ultrasonically derived arterial pressure in determination of amputation level. *Ann. Surg.* **41**, 731–7.

Dowd, G. S. E. (1982). Assessment of skin viability—with special reference to the transcutaneous oxygen monitor. MD thesis, Liverpool University.

—, Linge, K., and Bentley, G. (1983*a*). Measurement of transcutaneous oxygen pressure in normal and ischaemic skin. *J. Bone Jt Surg.* **65B**, 79–83.

— — — (1983*b*). The effect of age and sex of normal volunteers upon the transcutaneous oxygen tension in the lower limb. *Clin. Phys. Physiol. Meas.* **4**, 65–8.

Gerlach, J. V. (1851). Uber das Hautathmen. *Archs Anat. Physiol., Leipzig* **17**, 431–55.

Holloway, G. A. (1980). Cutaneous blood flow responses to injection trauma measured by laser-Doppler velocimetry. *J. invest. Derm.* **74**, 1–4.

Holstein, P., Nielsen, P. E., and Barras, J. P. (1979). Blood flow cessation at external pressure in the skin of normal human limbs. Photo-electric recordings compared to isotope washout and to local intraarterial blood pressure. *Acta orthopaed. scand.* **50**, 59–66.

—, Noer, I., Tonnesen, K. H., Sagar, P., and Lassen, N. A. (1978). Distal blood pressure in severe arterial insufficiency. In *Gangrene and severe ischaemia of the lower extremities* (ed. J. J. Bergan and J. S. T. Yao). Grune and Stratton, New York.

Horne, J. G. and Tanser, E. (1982). The assessment of skin viability in patients undergoing amputation for peripheral vascular disease using fluorescein angiography. *J. Bone Jt Surg.* **64A**, 260.

Huch, A., Huch, R., and Lübbers, D. W. (1969). Quantiative polarographic measurement of the oxygen pressure on the scalp in the new-born. *Arch. Gynak.* **207**, 443–51.

Keller, H. P., Klaue, P., and Lübbers, D. W. (1978). Transcutaneous pO_2 measurments for evaluating the oxygen supply of skin allo-autografts. *Eur. Surg. Res.* **10**, 272–8.

Mooney, V., Wagner, F. W., Waddel, J., and Acherson, T. (1976). The below-knee amputation for vascular disease. *J. Bone Jt Surg.* **58A**, 365–8.

Ohgi, S., Ito, K., and Mori, T. (1981). Quantitative evaluation of the skin circulation in ischeamic legs by $tcPO_2$ measurement. *Angiology* **32**, 833–9.

Peabody, J. L., Willis, M. M., Gregory, G. A., Tooley, W. H., and Lucey, J. F. (1978). Clinical limitations and advantages of transcutaneous oxygen electrodes. *Acta anaesth. scand.* **68**, 76–82.

Robinson, K. P. (1981). Surgical techniques in amputations in the dysvascular patient. Lecture given at the Royal College of Surgeons, London.

Rooth, G., Sjostedt, S., and Caligara, F. (1957). Bloodless determination of arterial oxygen tension by polarography. *Sci. Tools. LKB Inst. J.* **4**, 37–42.

Sarmiento, A., May, B. J., Sinclair, W. F., McCollough, N. C., and Williams, E. M. (1970). Lower extremity amputations. *Clin. Orthopaed. Rel. Res.* **68**, 22–31.

Siegel, M. E., Giragiana, F. A., White, R. I., Friedman, B. H., and Wagner, H. N. (1975). Peripheral vascular perfusion scanning, correlation with the arteriogram and clinical assessment in the patient with peripheral vascular disease. *Am. J. Roent. Radium. Ther. nucl. Med.* **125**, 628–33.

Waters, R. L., Perry, J., Antonelli, D., and Hislop, H. (1976). Energy cost of walking amputees: the influence of level of amputation. *J. Bone Jt Surg.* **58A**, 42–6.

Wyss, C. R., Matsen, F. A., King, R. V., Simmons, C. W., and Burgess, E. M. (1981). Dependence of $tcp0_2$ on local arterio-venous pressure gradient in normal subjects. *Clin. Sci.* **60**, 499–506.

Yao, S. T., Hobbs, J. T., and Irvine, W. T. (1969). Ankle systolic pressure measurements in arterial disease affecting the lower extremities. *Br. J. Surg.* **56**, 676–9.

20 The Measurement of Pressure Distribution on the Plantar Surface of Diabetic Feet

P. R. CAVANAGH, E. M. HENNIG, M. M. RODGERS, AND D. J. SANDERSON

INTRODUCTION

The medical complications of diabetes mellitus are often so overwhelming for the patient that they tend to dominate the early treatment of the disease. Frequently it is some years later, when the consequences of vascular insufficiency and neural degeneration become apparent, that attention is turned towards the orthopaedic problems of the feet. It is possible that occurrence of lesions on the feet of at least some diabetic patients is preventable if foot care is considered an important part of the total patient care during the early stages of the disease.

The cycle of neuropathy, angiopathy, ulcer formation and infection in the diabetic foot has been well described (Mooney and Wagner 1977). Excess local pressure on the plantar surface of the foot has long been implicated in the development of plantar keratosis and neuropathic lesions (Dickson and Diveley 1939) but only recently has evidence been gathered quantitatively to associate these factors (Stokes *et al.* 1975; Ctercteko *et al.* 1981).

Stokes *et al.* (1975) measured the load under the feet of diabetic and normal subjects using a force plate with an active area divided into 12 parallel beams, each 12.5 mm by 300 mm. Strain gauge methods were used to record the vertical component of load on each beam during the stance phase of gait so that either the mediolateral or anterioposterior distribution of foot pressure could be obtained. These authors demonstrated areas of heavy loading corresponding to the ulcer sites in the diabetic subjects with neuropathic lesions. Their subjects showed altered loading by a lateral shift of maximum local load on the forefoot and a decrease in the load carried by the toes.

A more recent study was conducted by Ctercteko *et al.* (1981) to examine the vertical forces under diabetic feet having neuropathic ulceration. The apparatus used was a load-sensitive surface divided into 128 load cells each 15 × 15 mm in area. Three groups were compared; (i) 24 diabetic patients with neuropathic ulceration of the foot; (ii) 21 diabetic patients with peripheral neuropathy but no ulceration; and (iii) normal subjects. The results agreed with the study by Stokes *et al.* (1975) in the reduced toe loading found in the diabetic patients, and in the correspondence between site of maximum force and ulcer site. The patients with neuropathic ulceration were significantly heavier than those without ulceration and, contrary to Stokes' finding, showed a tendency towards a medial shift of the peak forces. Ctercteko *et al.* also discussed the role of other factors—such as dryness, hyperkeratosis, muscular atrophy, and adipose tissue changes—in the aetiology of ulcer formation.

It is the purpose of the present chapter to introduce a new technique for the evaluation of plantar pressure distribution which has a greater spatial resolution than previous methods. A case is presented which illustrates the application of the device in the study of the diabetic foot.

METHODOLOGY

This study used a multielement piezoceramic platform to measure vertical pressure distribution under the foot (Fig. 20.1). Piezoelectricity is a phenomenon whereby electric dipoles are generated in certain anisotropic crystals, when subjected to mechanical stress. Piezoceramic materials are poled ceramics, which generate charge on their surface when subjected to mechanical stress. Piezoceramic materials can generate up to 100 times more electrical charge than the conventionally used quartz. This high charge output allows a considerable simplification of the necessary electronic processing. Low costs for transducer material and the electronics allow the construction of economical pressure

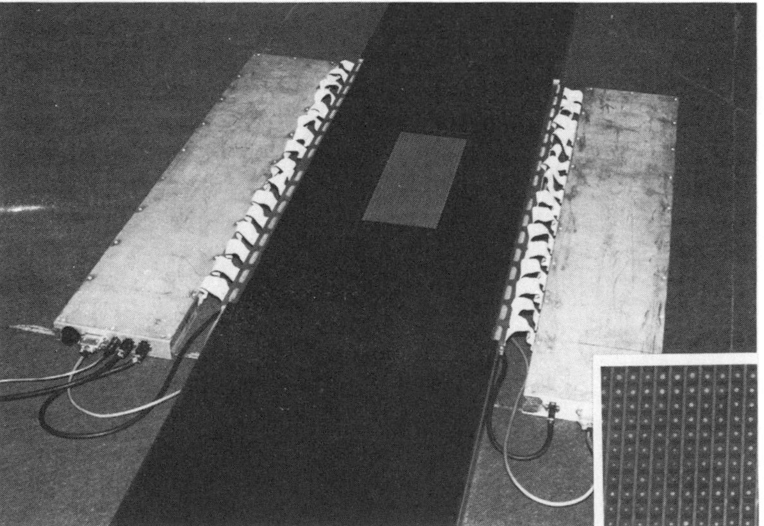

Fig. 20.1. The 1000-element piezoelectric platform. Amplifiers are in the rectangular cases on each side of the platform. *Inset* A close-up view of a section of the platform. Each element has dimensions 5 × 5 mm and the element spacing in 7.5 mm.

distribution devices with several hundred force transducers. A detailed description of the physical characteristics of piezoceramic force transducers, their use in pressure distribution devices, electronic processing, and data collection is given in Hennig *et al.* (1982).

Based on the same principles, Cavanagh and Hennig (1982) reported the construction of a rigid 1000-element piezoceramic pressure distribution platform. This platform was used in the present study. Its sensitive area was 15 × 37.5 cm, providing a spatial resolution of the pressure measurements of 7.5 × 7.5 mm. Individual charge amplifiers were used for each element and, by a multiplexing technique, a single analogue output line was created. An external memory and analogue-to-ditigal converter running under the control of an Apple II microcomputer were used to collect the data. A time resolution of approximately 30 ms between subsequent pressure measurements was used for data collection. A plot of individual force-time curves for selected elements could be obtained immediately (Fig. 20.2) but the data from all elements was routinely stored on floppy disks for offline transfer to a PDP 11/34 minicomputer and further numerical processing. A Megatek 7000 series graphics computer was then used to display a three-dimensional image of the pressure data.

The data presented here are part of a larger study in which 18 diabetic patients (age 25–73) served as subjects. Information regarding the history of foot

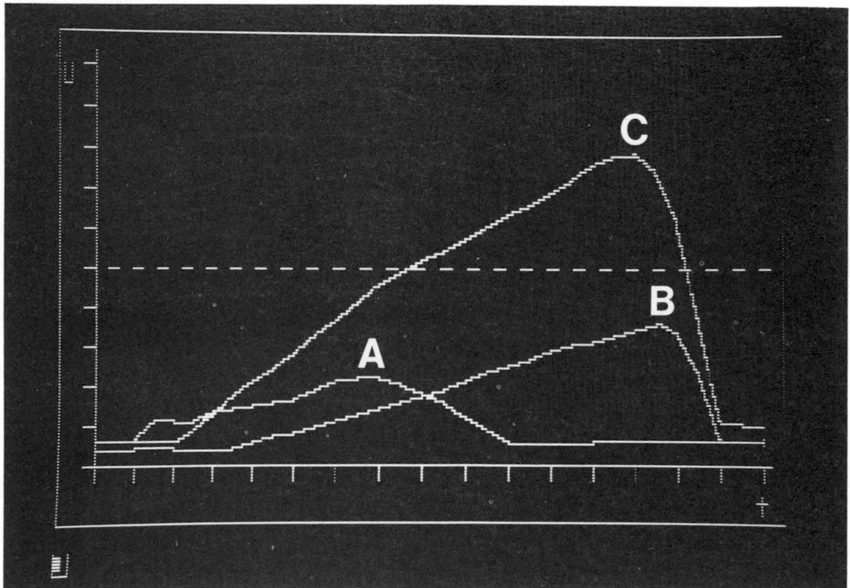

Fig. 20.2. Force–time curves from elements under the heel (A); medial forefoot (B); and lateral forefoot (C). The pattern from any element can be plotted immediately after data collection.

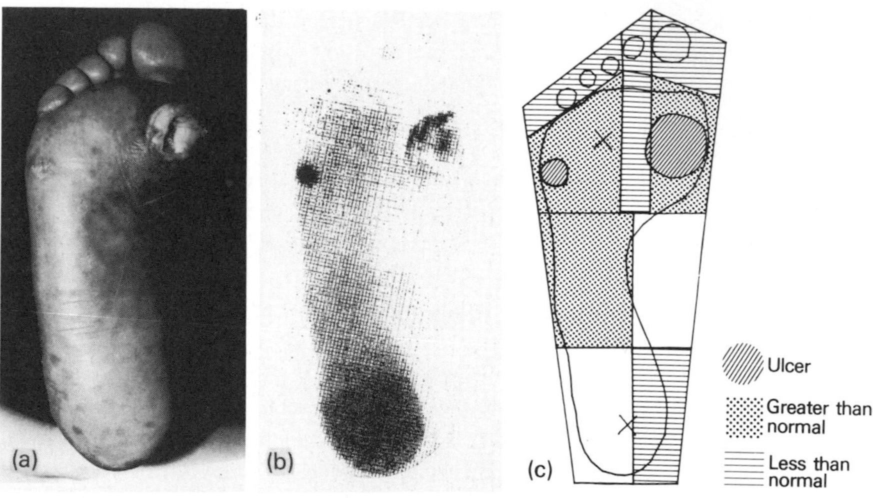

Fig. 20.3. (a) The plantar surface of the foot of the patient studied. In addition to the large open lesion under the first metatarsal head is a small healed lesion underneath the fifth metatarsal head. (b) A Harris mat print of the patient taken during walking. Notice the two areas of high pressure under the forefoot and the diffuse area under the rearfoot. (c) The result of impulse analysis showing regions of the foot which are loaded the same as (unshaded), less than (horizontal lines), and greater than (dotted) normal during the entire contact phase. The location of the lesions are also shown (diagonal lines).

problems was recorded from each patient. Following photography of the feet, Harris mat impressions which yielded a three-point scale of plantar pressures (Brand 1983) were made for comparison with the quantitative results (see Fig. 20.3).

To eliminate the difficult task of targeting on to the platform during walking, the patients made one step from standing on to the transducer area. Reference markers were placed on the foot and the location of ink impressions of these markers were used to describe foot placement on the platform for subsequent analysis. A standing footprint was also made on forensic paper to enable a numerical description of the foot outline to be made for association with the pressure distribution at the time of data display.

RESULTS

The patient was a male with a 19-year known history of diabetes mellitus. He presented with a healed ulcer under the fifth metatarsal head and an open lesion 4 cm in diameter under the first metatarsal head which had resisted healing for two years (Fig. 20.3(a)). Eight pressure distribution plots are shown in Fig. 20.4 with the times indicated in milliseconds from initial foot contact.

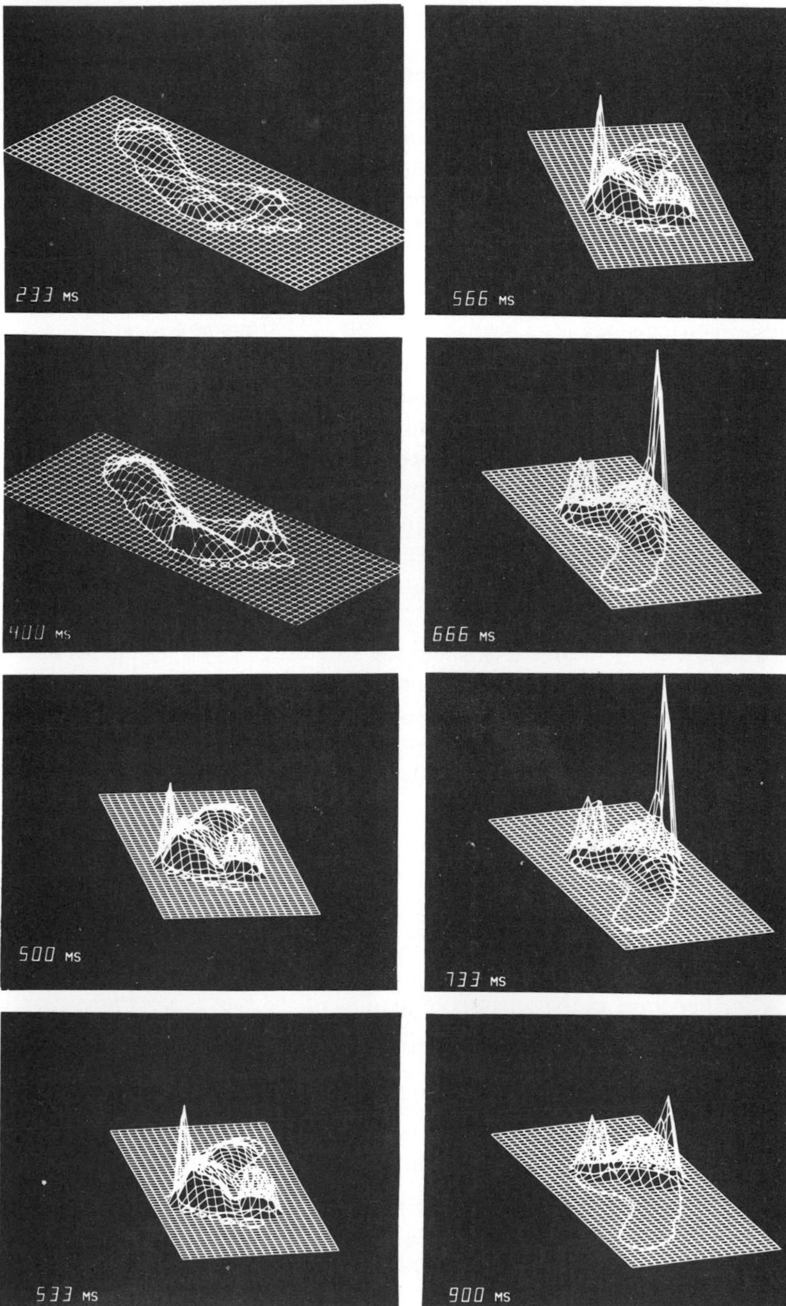

Fig. 20.4. Eight pressure distribution plots for various times after first contact (0 ms) of the foot with the platform. The foot outline is superimposed on the plot and the height of the graph above the plane of the platform is proportional to the pressure at that point. See text for further details.

These plots are individual frames from an animated motion picture which shows the temporal and spatial development of plantar pressures. Note that the orientation of the display is changed at 300 ms and again at 600 ms.

The patient made a flatfooted contact and by 233 ms exhibited a three-point contact with the heel, the first metatarsal head and a broad region of the lateral border of the foot all having approximately equal pressures. At 400 ms into the contact phase, there was a plateau of pressure along the length of the lateral border with the initiation of higher peaks under the first and fifth metatarsal heads. Heel pressures were slightly reduced at 500 ms and localized concentrations of pressures were very evident under the first and fifth metatarsal heads. The extremely localized pressure underneath the fifth continued to increase while the more extensive loading area under the first metatarsal head stayed at a fairly constant value until late support. The loading under the fifth showed approximately twice that of the first metatarsal head at 540 ms and almost three times the pressure by 660 ms. It is apparent that there is a 'valley' between the medial and lateral borders of the forefoot where no appreciable load was carried.

By the time the pressure under the fifth metatarsal head reached a peak value (733 ms) there was still no loading under any of the toes. From this point there was a steady decay of pressures under the forefoot. At 900 ms medial and lateral pressures were again similar and there was no evidence of involvement of the toes in the propulsive action.

It is clear from a comparison of Figs. 20.3(a) and 20.4 that the two sites of present and former lesions on the patient's foot correspond well with the regions of peak loading identified in the pressure distribution plots. However the higher loads are carried under the region where the previous lesion has undergone healing. Although the patient has diminished sensation, it is possible that he is modifying his gait to unload the region of greatest current concern. While this is allowing for some unloading of the open lesion it may result in exacerbation of the previous problem.

A further stage in the analysis of loading of the foot is a consideration of the total impulse sustained by the various anatomical regions. For this purpose, the digitized foot outline (Fig. 20.3(c)) has been divided into 10 distinct anatomical regions and the individual elements that lie within these regions have been automatically identified. Integrals of the force-time curves from the individual elements were calculated and summed to generate regional integrals (Clarke 1980). These impulse values represent the loading history that was experienced by the various regions throughout the contact phase. Comparison of the regional impulse values for this patient with 'normal' data are made in Fig. 20.3(c). Areas that are loaded less than, more than and the same as 'normal' are identified in the figure. The results present a simple summary of the loading history preserving the impressions of excessive medial and lateral forefoot loads and minimal toe involvement gained from the more detailed pressure distribution plots.

Examination of the Harris mat print (Fig. 20.3(b)) confirms the absence of toe loading and does enable us to identify three regions of high pressure. These include one diffuse area over the heel and two regions in the forefoot which apparently have similar peak pressures. Although the data from the platform show that, in fact, the magnitude of these pressures are very different, the Harris mat is nevertheless directing attention to the appropriate regions of the forefoot. It should not therefore be completely discarded in terms of clinical usefulness. However, its value as a quantitative indicator is clearly limited.

DISCUSSION

The graphic presentations of the data (Fig. 20.4) have considerable potential for use in a clinical setting. They contain a large amount of information (over 8000 data points) yet they present the data in such a form that is readily understandable and easily related tc the anatomy and pathology. The correspondence between the regions of maximum pressure and the sites of lesions are, in this patient, dramatic. This finding, also seen in many other patients, confirmed the results of previous investigators. This does not, however, preclude the importance of other factors (such as high shear stresses) in the aetiology of ulcer formation.

It is our feeling that the most useful application of this technique in the future is to patients who are in the early stages of the disease, who appear to be at high risk for ulcer development and yet exhibit no lesions at the time of study. Their classification of 'high risk' may be due to abnormal foot structure, atrophy of adipose tissue, significant plantar keratosis, radiographic evidence of joint degeneration, peripheral neuropathy or PVD. Once the regions of abnormal pressure distribution have been recognized, a variety of treatment regimes could be instituted to prevent breakdown of the foot. Pressure distribution measurement could also play a significant role in the evaluation of the success of the various treatments. It is our intention to investigate the use of pressure distribution measurement in such a prophylactic setting.

Acknowledgements

The generous co-operation of the members of the Department of Physical Therapy at the Union Memorial Hospital, Baltimore, MD is gratefully acknowledged.

REFERENCES

Brand, P. W. (1983). The diabetic foot. In *Diabetes mellitus: theory and practice*, 3rd edn (ed. M. Ellenberg and H. Rifkin). Medical Examination, New York.

Cavanagh, P. R. and Hennig, E. M. (1982). A new device for the measurement of pressure distribution on a rigid surface. *Med. Sci. Sport Exercise* **14**, 153.

Clarke, T. E. (1980). The pressure distribution under the foot during barefoot walking. Doctoral Dissertation, The Pennsylvania State University.

Ctercteko, G. C., Dhanendran, M., Hutton, W. C., and LeQuesne, L. P. (1981). Vertical forces acting on the feet of diabetic patients with neuropathic ulceration. *Br. J. Surg.* **68**, 608–14.

Dickson, F. D. and Diveley, R. L. (1939). *Functional disorders of the foot: diagnosis and treatment*. Lippincott, Philadelphia.

Hennig, E. M., Cavanagh, P. R., Albert, H. T., and Macmillan, N. H. (1982). A piezoelectric method of measuring the vertical contract stress beneath the human foot. *J. biomed. Engng* **4**, 213-22.

Mooney, V. and Wagner, F. W. (1977). Neurocirculatory disorders of the foot. *Clin. Orthopaed.* **122**, 53-61.

Stokes, I. A. F., Faris, I. B., and Hutton, W. C. (1975). The neuropathic ulcer and loads on the foot in diabetic patients. *Acta orthopaed. scand.* **46**, 839–47.

21 The Rheumatoid Foot During Gait

R. W. SOAMES, P. G. CARTER,
AND J. A. TOWLE

INTRODUCTION

Owing to its wide variety of functions the foot has been considered to be one of the most dynamic structures within the body. It provides physical contact with the environment, especially during gait when it must constantly adjust to the varying loads placed upon it during the initiation and termination of ground contact. Any changes in the structure and/or flexibility of the foot will modify its function, resulting in changes in the way in which the foot is used during gait.

In normal walking the heel strikes the ground first, followed by a rapid loading of the remainder of the foot. The centre of gravity of loading begins in the proximal part of the heel and passes over the medial side of the foot to the second metatarsal head, and ends at the lateral border of the great toe (Sharma *et al.* 1979). As the load moves forward and approaches the metatarsal heads, the heel leaves the ground, body weight being borne entirely by the forefoot. Although the highest loads are exerted on the heel, it is the forefoot that is involved for the greater part of foot contact time (Grundy *et al.* 1975). The medial side of the forefoot is heavily loaded, with the hallux carrying about twice as much as the metatarsal head (Dhanendran *et al.* 1980).

The toes serve mainly as an accessory to the ball of the foot, providing attachment of the long flexor tendons. By differential contraction of these flexors it is possible to adjust the distribution of pressure between points on the ball of the foot. Because of the strength of the great toe and the long flexor tendons attached to it, this part of the foot is usually the last to leave the ground, contributing the final touch to the control of movement. Up to 20 per cent of body weight acts on the great toe at toe-off (Sharma *et al.* 1979). According to Stokes *et al.* (1979) this is counteracted by tension in the flexor tendons, which react with the force on the great toe to produce a resultant force approaching body weight on the first metatarsophalangeal joint. It would not therefore be surprising to find that in subjects with rheumatoid arthritis that there is an abnormal loading of this joint.

The commonest site for involvement of the foot by rheumatoid arthritis is in the metatarsophalangeal joints. According to Calabro (1962) and Dixon (1970) the subsequent pattern of events follows a common pathway culminating in the metatarsal heads becoming subcutaneous on the sole of the foot. The metatarsal heads then sustain more of the weight-bearing function, the toes being inactive, and this can lead to metatarsal fracture. Collis and Jayson (1972) and Soames (1983), among others, have observed that the greatest pressures in the forefoot are beneath the second and third metatarsal heads. The development of pressure lesions under the second and third metatarsal heads is therefore to be expected, and has been observed in 40 per cent of patients by Vidigal *et al.* (1975). This course of events may be modified by the patients' own footwear. Distorted packing of the toes will influence the position of the great toe, which often assumes a valgus position. There may also be dislocation of the sesamoid bones in flexor hallucis brevis, which become repositioned in the first web space and as such take no part in weight-bearing (Vidigal *et al.* 1975).

According to Vidigal *et al.* the hindfoot is rarely involved in the absence of midtarsal and metatarsophalangeal joint disease. Similarly subtalar joint disease rarely occurs without midtarsal and metatarsophalangeal joint disease. Vainio (1956) has stressed the importance of the talonavicular and naviculocuneiform joints, pointing out that damage to them through rheumatoid arthritis causes the longitudinal arch of the foot to be disturbed. The foot then becomes flattened and subsequently adopts a valgus position through the weight-bearing forces.

Depression of the metatarsal heads in rheumatoid arthritis, particularly the third, produces a plantarward convex arch. This may be the result of pressure transmitted by the near vertical proximal phalanx after dorsal dislocation of the metatarsophalangeal joint (Craxford *et al.* 1982). Furthermore, bone erosion from the condyles of the metatarsal heads, owing to rheumatoid granulation tissue, will sharpen and decrease the weight-bearing bone surface.

Although the causes of pain in any one subject are likely to be variable and multiple, their effects on the stance phase of gait will be similar in that the normal rolling action of the foot is replaced by a flat-footed antalgic type of gait (Craxford *et al.* 1982). Thompson (1964) has likened the foot in these circumstances to a pedestal rather than a lever.

In a qualitative study Barrett (1976) observed that in patients with rheumatoid arthritis there are high pressures under the second and third metatarsal heads. Sharma *et al.* (1979) reported a slight shifting of loading away from the medial side of the forefoot to the lateral side in patients with rheumatoid arthritis; only the first metatarsal head showed any significant change in loading. These same patients, however, showed a significant decrease in the weight-bearing function of the toes.

The present study was undertaken to determine the function of the rheumatoid foot during the stance phase of gait, in terms of both the plantar pedal pressure distribution and the timing of foot contact. In addition the influence of the patient's footwear on these parameters was of interest.

METHOD

Twenty-five patients (nine male, 16 female) with rheumatoid arthritis of the foot and ankle were drawn from those attending the Rheumatology Outpatient's Clinic at Westminster Hospital, London. Each patient was examined by senior hospital staff, who assessed the extent and severity of the condition in each foot and ankle. From this examination and from the patient's history four feet were rejected from this study because of surgical intervention of the forefoot, leaving a total of 46 feet. Of these 11 feet (six right and five left) had erosive changes involving one or more metatarsophalangeal joints and were deemed to be one group (Group E). The remaining 35 feet (18 right and 17 left) showed no erosive changes and were thus considered as a second group of patients (Group RA). In this second group there was arthritic involvement of the knee of the same side for 16 of the feet. For comparison, recordings were taken from 10 subjects (seven male, three female) with no known history of rheumatoid arthritis and with no apparent disorders of gait.

The plantar pedal pressures at selected sites on the sole of the foot (Fig. 21.1) were recorded using small semiconductor strain gauge transducers, the outputs of which were led to a PDP-11 computer, digitized and stored on magnetic tape for later analysis. Full details are given in Soames *et al.* (1982).

Recordings were taken from each foot, both with the subject walking barefoot and when wearing normal footwear, which usually meant the most comfortable pair of shoes they had. Footwear was therefore not uniform. Each recording comprised between six and eight consecutive steps. From these recordings peak pressure, transducer contact time as a percentage of total foot contact, and time to peak pressure as a percentage of transducer contact time were determined for each subject for both barefoot and shod conditions.

RESULTS

The age and weight characteristics of each group of subjects are presented in Table 21.1. As can be seen the non-arthritic group are significantly younger than both patient groups. Since there is no information available which suggests that age has an effect on pedal plantar pressures and their distribution, the age difference between the groups was not deemed to influence the findings. There is no significant difference between the mean weights of the normal subjects and group RA. However, both groups are significantly lighter than group E. Again, this is not likely to have any influence on the pedal pressures as Soames (1983) has found no relationship between body weight and the pressures under the foot for normal subjects.

All differences mentioned are significant at the 1 per cent level.

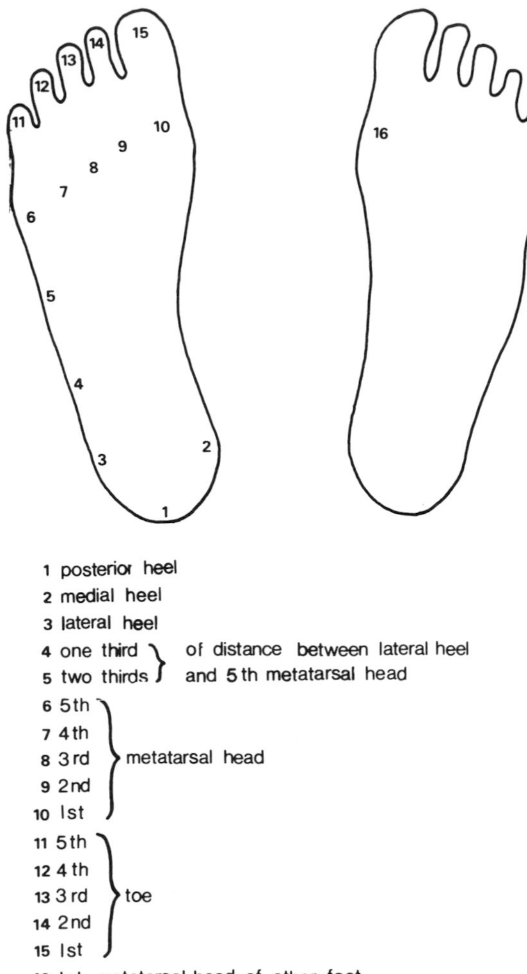

Fig. 21.1. Locations of the transducers on the sole of the foot.

Table 21.1. *The means and standard deviations of the age and weight of the non-arthritic subjects and group RA and group E of the patients*

	Age $\bar{x}$ (SD)	Weight $\bar{x}$ (SD)
Non-arthritic	22.3 (2.3)	64.1 (5.7)
Group RA	55.4 (11.9)	63.7 (9.7)
Group E	50.7 (13.8)	76.4 (15.6)

Peak pressures

The mean peak pressures (expressed as kPa × 100), together with their associated standard deviations, are presented in Fig. 21.2 for the non-arthritic subjects and groups RA and E of the patients, for both barefoot walking and wearing shoes. For barefoot walking there are significant differences in the pressure distribution under the foot between the non-arthritic subjects and both patient groups. There are also some significant differences between both patient groups, particularly under the metatarsal heads. Pressure under the heel shows significant differences between normal subjects and both patient groups. The pattern of pressures across the metatarsal heads in the non-arthritic subjects is similar to that observed previously (Soames, in preparation); viz. the highest pressures are observed under the third metatarsal head, with a gradual fall-off both medially and laterally. The only significant difference between group RA and the normals

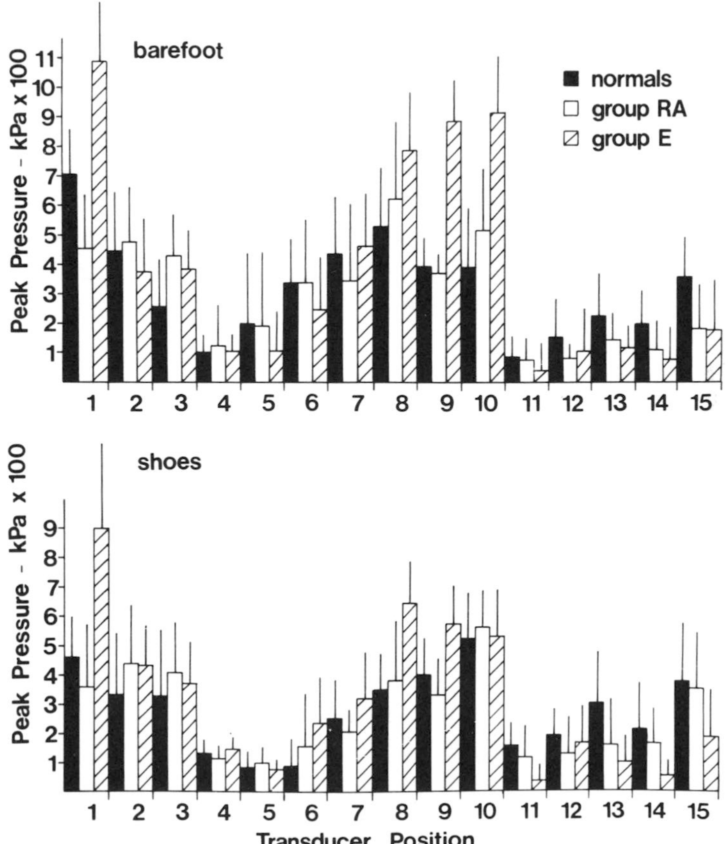

Fig. 21.2. The mean peak pressures (kPa × 100), and associated standard deviations for each transducer location.

is in the increased peak pressure under the first metatarsal head. In group E peak pressure under the medial three metatarsal heads are all significantly greater than those observed in the non-arthritic group. Apart from the fifth, peak pressure under the metarsal heads are significantly greater in group E than in group RA. Peak pressures beneath the toes in both arthritic groups are generally significantly lower than normal, there being no significant difference between the two arthritic groups.

In normal subjects, wearing shoes tends to spread the load more evenly under the heel and changes the pattern of pressures under the metatarsal heads, with the greatest pressure now being associated with the first metatarsal and gradually decreasing to the fifth. The pressure distribution beneath the toes is similar to that when walking barefoot.

The influence of shoes on foot pressure patterns in each of the arthritic groups is different, particularly with respect to the metatarsals and toes. In group E there is a significant decrease in pressures under the medial four metatarsals, with no real change being observed under the toes. In group RA there are significant decreases in pressure under the lateral four metatarsals, which is associated with significant increases in pressures under the toes.

Contact times (as percentage of total foot contact time)

Mean contact times, as a percentage of total foot contact time, and their associated standard deviations are shown for all groups in Fig. 21.3, for both wearing shoes and walking barefoot. The time for which the heel is in contact with the ground during the stance phase of gait is significantly increased for both groups of patients compared with the normal subjects, both when wearing shoes and when walking barefoot (Fig. 21.3). In barefoot walking no differences in contact times under the metatarsal heads between the normal subjects and both arthritic groups were observed. However, when wearing shoes contact times for both arthritic groups are significantly greater than normal for the third and fourth metatarsal heads, and significantly lower under the first metatarsal. Only the second and fifth metatarsals show any significant differences in contact times between the two arthritic groups.

Contact times for the toes show significant differences between all three groups of subjects for both walking barefoot and wearing shoes. In barefoot walking, contact times under the medial three toes are significantly lower than normal in group RA. This is in contrast to group E whose contact times under the second to fourth toes are significantly greater than normal, and only that under the great toe is lower than normal. There are some significant differences between the two arthritic groups. When wearing shoes the contact times under all toes are significantly greater in group E than in group RA, each group having a different pattern of contact times from those associated with normal subjects.

For the normal subjects, wearing shoes significantly increases the contact

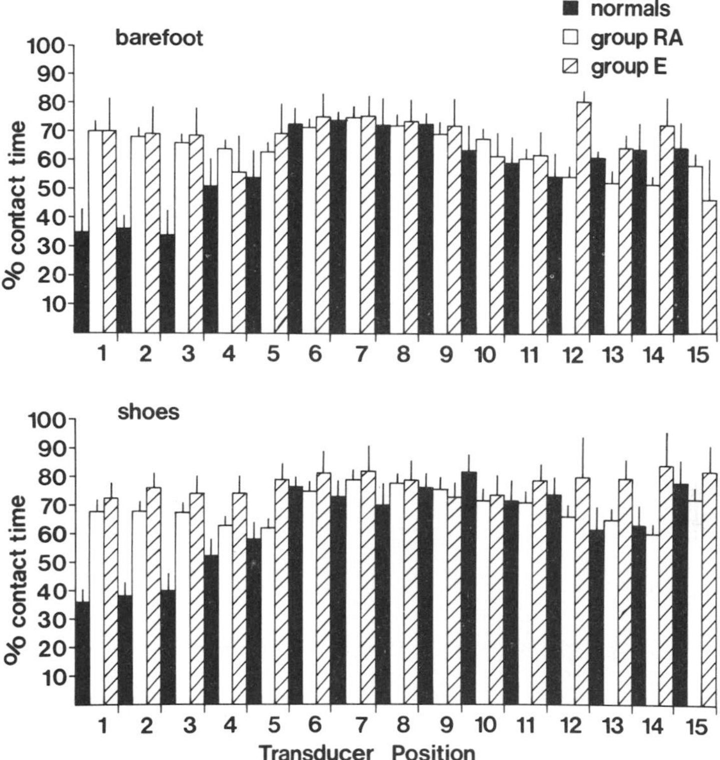

Fig. 21.3. The means and standard deviations of foot contact times (expressed as a percentage of total foot contact), for each transducer location.

times under the lateral heel; first, second, and fifth metatarsal heads; and the first, fourth, and fifth toes. In group RA there are significant increases in contact times under all metatarsal heads and toes except the fifth, which shows a significant decrease when wearing shoes. There are significant increases in contact times under the posterior heel, first, third, and fourth metatarsal heads, and all but the fourth toe in group E when wearing shoes.

It appears that in both arthritic groups the normal rolling action of the foot is not present, instead a rather 'flat-footed' placement of the foot on the ground is seen. In both of these groups, however, the general influence of shoes is to increase the time for which the foot is in contact with the ground.

Rise time to peak pressure (as percentage of transducer contact time)

The mean time taken to reach peak pressure from initial transducer contact, i.e. rise time, expressed as a percentage of the contact time for that transducer

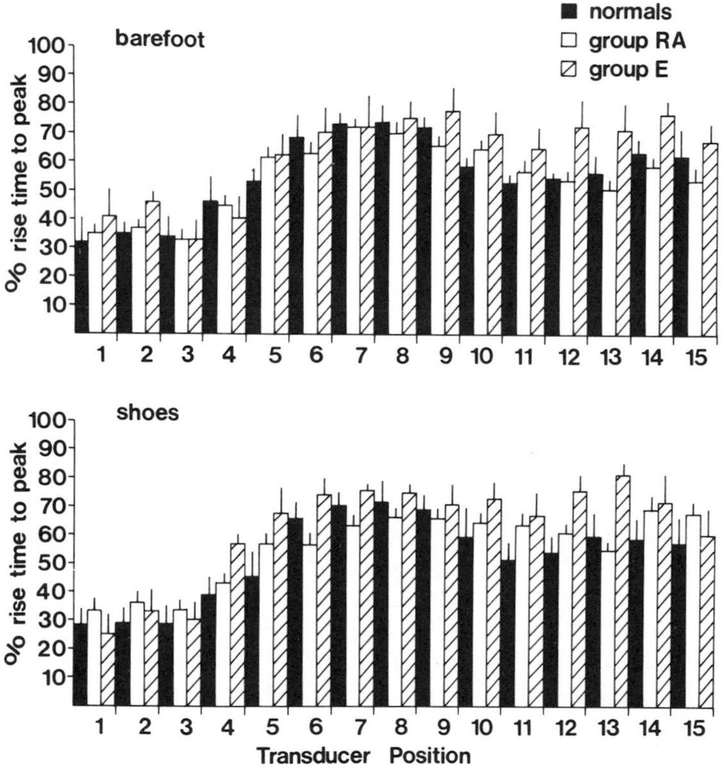

Fig. 21.4. The means and standard deviations of the rise time to peak pressure (expressed as a percentage of transducer contact time), for each transducer location.

site, are plotted in Fig. 21.4 for all groups. These rise times show significant differences between all three groups under the heel for barefoot walking, the shortest rise time being in the non-arthritic subjects, then group RA and finally group E. However, when wearing shoes the differences between group E and the normals disappear. In both of these groups rise time to peak pressure is significantly shorter than in group RA.

The rise times to peak pressure under all except the fourth metatarsal are significantly longer in group E than in group RA. Both arthritic groups have increased rise times under the first metatarsal compared with normal. However, group RA have significantly reduced rise times under most of the remaining metatarsals, while group E also have an increased rise time under the second metatarsal. There are some differences between group E and the normals in rise times under the metatarsals when wearing shoes, those under the first, fourth, and fifth metatarsals being significantly shorter in the non-arthritic group.

The rise times under all of the toes in group E are significantly longer than normal and those observed in group RA. In general the shortest times to peak pressure under the toes when walking barefoot are associated with group RA. With the exception of the great toe in group E and the third toe in group RA, the rise times in both arthritic groups are significantly longer than normal when wearing shoes.

The effect of wearing shoes for the non-arthritic subjects is to significantly lengthen rise times under the heel and midfoot region. However, for group RA rise times under the heel and all of the toes are significantly lengthened, while those under the midfoot and lateral three metatarsal heads are significantly reduced. In contrast, the influence of shoes in group E is to significantly reduce rise times under the heel, as well as under the second metatarsal head and great toe.

DISCUSSION

The patterns of pressure observed under the foot in both arthritic groups are different in many respects from those recorded in the non-arthritic group. The greatest difference was seen in the group of patients who had erosive changes involving the metatarsophalageal joints. The only part of the foot which does not show any change in this group is the midfoot region. From the mean peak pressure data it would appear that the medial metatarsals have assumed an increased role in weight bearing in the stance phase of gait in the group E patients, both when wearing shoes and when walking barefoot. This is at variance with the findings of Sharma *et al.* (1979), who found a significant decrease in loading under the first metatarsal head in similar subjects. However, in agreement with Sharma *et al.*, loading under the toes in both arthritic groups is significantly reduced.

Peak pressures under both the first and second metatarsal heads in the patients with erosive changes are almost twice as great as corresponding pressures both in normal subjects and the other arthritic group. In rheumatoid arthritis there are inequalities in the strength and direction of pull of the long flexor and extensor tendons, which tend to cause an anterior displacement of the fat pads beneath the metatarsal heads, leading to a decrease in the cushioning function of these fat pads. Where there is erosion of the condyles of the metatarsal head, the weight-bearing surface of the bone will be decreased, and the pressures recorded are likely to be significantly greater than those in a non-arthritic population.

In many respects the normal foot functions as an arch, whose integrity is maintained by the plantar aponeurosis and long flexor tendons. The action of body weight is to flatten the arch thus creating tension in the aponeurosis and flexor tendons, causing a plantar flexion of the toes so that they press firmly on the ground. In this way the area of the base of support is extended forwards and the pads of the toes become weight-bearing. However, in many cases of rheumatoid arthritis there is a dorsal contracture leading to an extension of the toes. As a result the share of body weight taken by the toes is reduced and the metatarsal heads therefore assume an increased load.

Extension of the toes is also in keeping with the observation that pressures beneath the toes in the arthritic subjects are lower than those seen in normal subjects. In fact the greatest differences in pressures under the toes, particularly between patients with erosive changes and normal subjects, are found under those toes whose metatarsal heads showed significantly increased pressures.

The duration of pressures under the heel in the arthritic subjects are twice as long as those seen in normal subjects. There may be two possible explanations for this: (i) it may be an attempt by the patient to shift some of the increased load from the metatarsal heads to other parts of the foot; or (ii) it may be an adjustment by the patient to reduce pain in the forefoot during gait.

From the foot contact data it is apparent that the normal rolling action of the foot during stance is not present in the arthritic subjects. Instead the foot is 'placed' on the ground by a downward movement of the whole foot, and at the end of stance it is 'lifted' off the ground. This type of action would account for the increased contact times observed under the heel. Furthermore, owing to the dorsiflexion contracture of the toes, the push-off phase of gait will be modified. In the normal situation as body weight moves over the metatarsal heads the toes press firmly against the ground ready for push-off. This action will be severely modified in the patients and the levering action of the foot will be impeded. As Thompson (1964) pointed out, in these conditions the foot is being used as a pedestal rather than as a lever.

In the patients there is probably arthritic involvement of the midtarsal joint, which affects the normal function of the foot during stance, where it allows the forefoot to remain on the ground while the hindfoot leaves the ground. The normal midtarsal joint plays an important role in inversion and eversion of the foot. In the early part of stance the axes of its two component joints are almost parallel, so that mobility is high and stability low. In the latter part of stance the two component joints are no longer parallel and mobility is lost in favour of stability: this stability is necessary for efficient push-off. If the midtarsal joint has undergone arthritic change then there will be decreased mobility and stability, thus influencing foot function. Perhaps placing the whole foot on and off the ground is a way of achieving a compromise with respect to these two functions.

A further and important factor regarding the mode of foot placement in the patients is concerned with balance. If the toes are dorsiflexed the area of the base of support is effectively reduced. It may be that placing the whole foot on the ground is a compensation for the absence of toe function, and helps the patient to maintain balance during gait. Finally, it must be recognized that the patients are walking in such a way so as to reduce pain under specific areas of the foot, and thus enable them to be mobile for longer periods of time.

SUMMARY

The spatial and temporal distribution of plantar pedal pressures during gait have been studied in two groups of arthritic subjects, one of whom had erosive changes involving one or more metatarsophalangeal joints. When walking barefoot there are significant differences in the pressure distribution under the foot between the arthritic and normal subjects. There are also some significant differences between the two arthritic groups, particularly under the metatarsal heads. In general, wearing shoes tends to spread the pressures more evenly under the heel, reduces pressures under the metatarsal heads, and increases pressures under the toes in the arthritic subjects.

The time for which the heel is in contact with the ground in the arthritic subjects is twice as long as for normal subjects, both when wearing shoes and when walking barefoot. It would appear that in the patients the normal rolling action of the foot is absent; instead the foot is used more as a pedestal than a lever. The main differences in rise times to peak pressure are found under the toes where the generation of peak pressure is significantly prolonged for both walking conditions, particularly in those with erosive changes.

It is suggested that one reason why the foot may be used as a pedestal is that in the absence of support from the toes, and a consequent reduction in the area of the base of support, it helps the patient maintain balance during gait.

Acknowledgement

This work has been carried out with the support of the Arthritis and Rheumatism Council of Great Britain.

REFERENCES

Barrett, J. P. (1976). Plantar pressure measurements, rational shoe wear in patients with rheumatoid arthritis. *J. Am. med. Ass.* **235**, 1138–9.

Calabro, J. J. (1962). A critical evaluation of the diagnostic features of the feet in rheumatoid arthritis. *Arthritis Rheum.* **5**, 19–29.

Collis, W. J. M. F. and Jayson, M. I. V. (1972). Measurement of pedal pressures: an illustration of a method. *Annls rheum. Dis.* **31**, 215–17.

Craxford, A. D., Stevens, J., and Park, C. (1982). Management of the deformed rheumatoid foot. *Clin. Orthopaed. related Res.* **166**, 121–6.

Dhanendran, M., Pollard, J. P., and Hutton, W. C. (1980). Mechanics of the hallux valgus foot and the effects of Keller's operation. *Acta orthopaed. scand.* **51**, 1007–12.

Dixon, A. St. J. (1970). Medical aspects of the rheumatoid foot. *Proc. R. Soc. Med.* **63**, 677–9.

Grundy, M., Tosh, P. A., McLeish, R. D., and Smidt, L. (1975). An investigation of the centres of pressures under the foot whilst walking. *J. Bone Jt Surg.* **57B**, 98–103.

Sharma, M., Dhanendran, M., Hutton, W. C., and Corbett, M. (1979). Changes in load bearing in the rheumatoid foot. *Annls rheum. Dis.* **38**, 549–52.

Soames, R. W. Foot pressure patterns during gait. In preparation.
—, Stott, J. R. R., Goodbody, A., Blake, C. D., and Brewerton, D. A. (1982). Measurement of pressure under the foot during function. *Med. biol. Engng Comput.* **20**, 489–95.
Stokes, I. A. F., Hutton, W. C., and Stott, J. R. R. (1979). Forces acting under the metatarsals during normal walking. *J. Anat.* **129**, 579–90.
Thompson, T. C. (1964). Surgical treatment of disorders of the forepart of the foot. *J. Bone Jt Surg.* **46A**, 1117–20.
Vainio, K. (1956). The rheumatic foot: a clinical study with pathological and roentgenological comments. *Annales Chir. Gynaec. fenn.* **45**, Suppl. 1.
Vidigal, E., Jacoby, R. K., Dixon, A. St. J., Ratliff, A. H., and Kirkup, J. (1975). The foot in chronic rheumatoid arthritis. *Annls rheum. Dis.* **34**, 292–7.

22 The Contribution of Skin, Fascia, and Ligaments to Resisting Flexion of the Lumbar Spine

K. M. TESH, J. H. EVANS, J. SHAW DUNN, AND J. P. O'BRIEN

INTRODUCTION

Motion of the lumbar spine is determined by the interaction of muscle and connective tissue. The effectiveness of these tissues in controlling spinal motion is dependent on their effective lever arms at the points of attachment. This study will investigate the contribution of skin and the thoracolumbar fascia which, being the most dorsal structures, are in an advantageous position to resist flexion of the spine.

CONTRIBUTION OF THE THORACOLUMBAR FASCIA AND POSTERIOR LIGAMENTS

For many years it has been accepted that muscle contraction of the abdominal wall can generate an intra-abdominal pressure which may stabilize the spinal column during flexion. That the tensed abdomen can act as a support is understood if one considers a tube-shaped balloon wedged between the pelvic floor and the diaphragm in the abdominal cavity (Bartelink 1957). The circumferential component of muscle action in the abdominal wall would squeeze the balloon, reducing its diameter, but simultaneously bulging the ends outward. The thrust on the diaphragm would be transmitted to the vertebral column via the rib cage. This cephalocaudal force will reduce the compressive load on the lumbar spine and, as it acts anterior to the column, will produce an extensor moment.

In 1980 Fairbank and O'Brien carried out an experiment to simulate abdominal muscle action by pulling on the lateral margins of the thoracolumbar fascia. Assuming a distance of 4 cm from the tip of the posterior process to the centre of rotation, the lumbar spine extended 4° from the neutral position in the sagittal plane. The extension movement would increase the axial compressive load on the intervertebral disc.

Although both mechanisms described above produce an extensor moment on the vertebral column the resultant axial loads oppose one another.

To explain the mechanism by which the thoracolumbar fascia extends the spine, Fairbank and O'Brien proposed that the ligamentous sheet interacted with the vertebral ligaments. The fibre orientation of the ligaments was thought to be such that extension of the lumbar spine could be achieved when acted upon by the thoracolumbar fascia. This mechanism is in addition to the one proposed by Farfan (1973) which involved the thoracolumbar fascia deforming like a net under tension.

The anatomy of the thoracolumbar fascia and its attachment to the spinal column have not been described in sufficient detail to enable the various mechanisms to be tested. In view of the potential importance of the thoracolumbar fascia in spinal mechanics a detailed study of the anatomy and microstructure of the fascia and its interaction with the vertebral ligaments has been undertaken.

SOFT TISSUE MORPHOLOGY

Dissections were carried out on fresh and embalmed cadaveric specimens in the lumbar region to reveal the three-dimensional anatomy of the following regions:

- The posterior layer of the thoracolumbar fascia.
- The junction at the midline of the posterior fascia layer with the supraspinous and bifid interspinous ligament. (To study this region a transverse section through the interspinous space was made.)
- The middle layer of the thoracolumbar fascia at its attachment to the transverse processes.

The microstructure of the posterior thoracolumbar fascia layer was studied using polarized light microscopy and conventional histology on full-thickness fascia samples. Histological sections were made to analyse the midline junction of the fascia layer and vertebral ligaments. Van Giesons solution was used to stain collagen fibres red. Weigert's iron haemotoxylin solution was used to stain the nuclei of the fibroblasts.

The posterior layer of the thoracolumbar fascia is composed of three laminae. Fibre direction alternates between laminae providing a network structure. Each lamina of the fascia displays collagen fibres with their typical rippled appearance. The middle lamina is the major part of the posterior layer, having a thick, closely packed, parallel array of collagen fibres. A sagittal section close to the midline reveals that in the thick midlamina the collagen fibres are arranged principally in a circumferential direction. The fibres in the outer laminae of the fascia sandwich run predominantly longitudinally. There is some interweaving of fibres from the superficial and deep laminae piercing the midlamina at the lamina junctions (Fig. 22.1).

The three laminae of the posterior thoracolumbar fascia layer terminate at different sites in the midline (Fig. 22.2). The outer lamina runs into the superficial fibres of the supraspinous ligament. The middle lamina runs deeper along

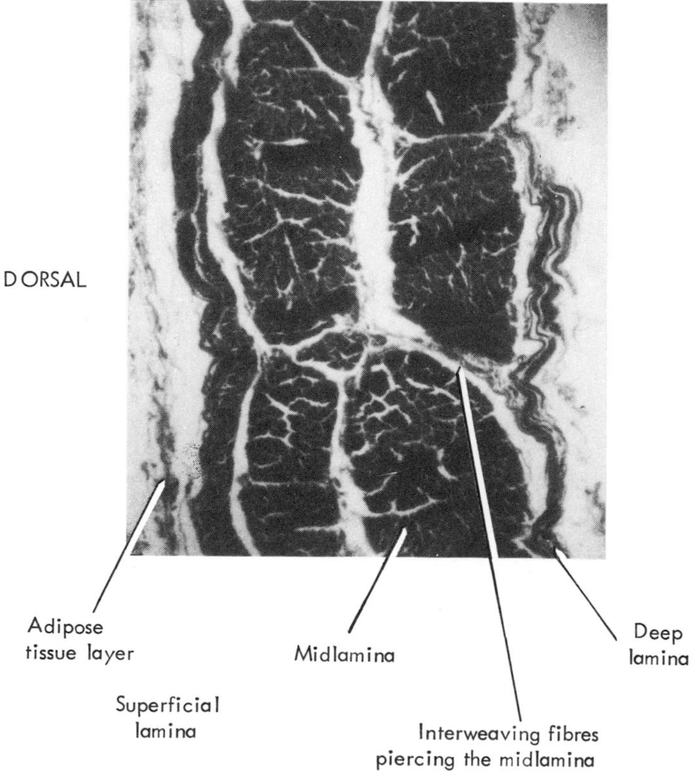

Fig. 22.1. A full-thickness sagittal section of the posterior thoracolumbar fascia. (X 160.) Specimen taken 3 cm from the midline at the L3–4 level.

the side of the ligament and then splits into two branches, one branch running medially into the deep part of the supraspinous ligament, while the other branch continues anteriorly towards the interspinous ligament. The inner lamina runs down the outside of the midline soft tissues, passing the supraspinous ligament, and merging with the interspinous ligament. The direction of the fibres in all three laminae are running predominantly in an anterior-posterior direction.

The middle layer of the thoracolumbar fascia terminates and anchors firmly onto the tips of the transverse processes and is composed of two sets of fibres which radiate from the anchor points. At these force-transmission sites the connective tissue layer was markedly more thick and dense. The middle layer arches between the tips of the transverse processes with no substantial attachment to the inter-transverse ligament.

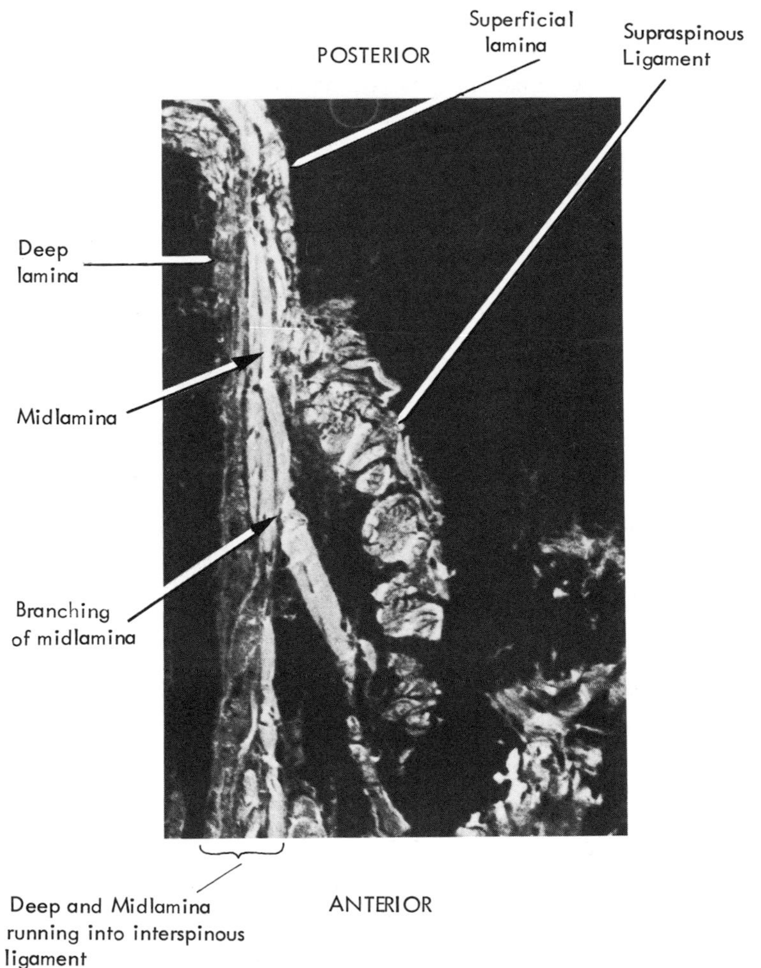

Fig. 22.2. A transverse section through the interspinous space at the L3–4 level showing the three laminae of the posterior thoracolumbar fascia entering the midline. (X 16.)

DISCUSSION

The interweaving arrangement between laminae in the posterior layer of the thoracolumbar fascia gives strong evidence for a net-type deformable ligamentous sheet. In 1978, Bazergui *et al.* performed mechanical tests on whole sheets of thoracolumbar fascia. The ligamentous sheet was pulled longitudinally to simulate elongation during flexion manœuvres. At the highest loads, longitudinal extensions of 8 per cent produced lateral contractions of 13 per cent in the ligamentous sheet. The ratio of contraction to extension is typical of a net-like structure.

The thoracolumbar fascia could be expected to resist forward bending by its relatively inextensible nature during loading. A circumferential force at the lateral margins of the fascia sheet set up by abdominal muscle tension could maintain its width thus rendering it stiffer in extension and could directly create a tension along the axis of the spine. The attachment of the fascia to the tips of the posterior spinous processes enables such a longitudinal force to be transmitted directly to the lumbar spine, thus drawing together the processes and extending the spine.

Microscopic examination of the midline ligamentous structures revealed that the posterior layer of the thoracolumbar fascia interacts with the supraspinous and interspinous ligaments. The superficial and middle laminae of the fascia merge with both ligaments, whereas the deep lamina interacts with the interspinous ligament only. The anterior-posterior orientation of collagen fibres in the fascia layer would suggest that tensing of the abdominal muscles would 'bow-string' the vertebral ligaments posteriorly, drawing the processes together and bringing the lumbar spine into extension. Fairbank and O'Brien (1980) suggested that the thoracolumbar fascia attached to the bifid interspinous ligament and caused an extension of the lumbar spine by pulling laterally on the lips of the ligament. Our findings have demonstrated that the fascia interacts with both the supraspinous and interspinous ligaments. Extension of the lumbar spine can still be produced by tension in the fascia directed more posteriorly.

It is interesting to note that below the level of the posterior spinous process of L4 the superficial fibres of the thoracolumbar fascia decussate the midline without any attachment to the midline ligamentous structures. Heylings (1978) showed that the supraspinous ligament terminated at the L4–5 interspace and its absence may account for the change in the anatomy of the fascia below this level.

The middle layer of the thoracolumbar fascia attaches directly to the vertebral column without involving the intertransverse ligaments. This finding conflicts with the ideas of Fairbank and O'Brien (1980) who suggested that the middle layer of the thoracolumbar fascia could stabilize the lumbar spine by maintaining a tension in the intertransverse ligament. However, the absence of a ligament-fascia interaction does not detract from the stabilizing potential of the middle layer. The attachment of the middle layer to the tips of the transverse processes is posterior to the centre of rotation in the sagittal plane of motion. Hence a lateral pull on the arcading layer resulting from abdominal muscle action would develop an axial component of force capable of extending the lumbar spine. The anchorage of the middle fascial layer to the tips of the transverse processes may well help to stabilize the spine during flexion as well as lateral bending in the coronal plane.

The attachment points of the posterior and middle layer of the thoracolumbar fascia allow it to act on a considerable lever arm. Large moments can be generated about the vertebral column when the fascia is put under tension.

The aponeurosis of the latissimus dorsi muscle constitutes a part of the posterior layer of the thoracolumbar fascia. In electromyographic studies carried out by Farfan (1977) in the compilation of his computer model of the lumbar spine, active potentials were measured in the latissimus dorsi muscle when the arms were tensed. During lifting manœuvres the arms will be tensed as the inertia of the weight is overcome; an oblique force will be transmitted to the fascia sheet via the aponeurosis of the latissimus dorsi muscle. During lifting the thoracolumbar fascia sheet would be loaded in three directions: longitudinally from increased convexity of the spine, circumferentially from abdominal muscle action, and obliquely (60° to the midline) from the latissimus dorsi muscle. The behaviour of the ligamentous sheet under triaxial loading is likely to be markedly different from that detailed above under biaxial loading.

CONTRIBUTION OF SKIN

The supraspinous ligament is usually considered to be the most dorsal structure resisting extreme flexion of the spine. However, skin, being furthest from the bending axis, can contribute significantly to balancing this bending moment, even at moderate stress levels. An investigation was undertaken to measure the contribution of the skin in resisting forward bending under normal body weight.

The load-strain characteristic for skin in six young, normal individuals was determined at the L4-5 level. With a subject lying in the prone position with the head turned to one side a skin tensometer (Gibson *et al.* 1969) was used to stretch the skin, 2 cm lateral to the midline and in a caudocranial direction. Two adhesive tabs were attached to the skin surface and were drawn apart by the tensometer device. The movement of the tabs, initially 3 cm apart, was recorded against applied force. It was not possible to load the skin directly over the midline because the protuding posterior spinous processes did not allow the tabs to sit flat on the skin surface.

On obtaining a subject's skin characteristics, strains were measured on the same individual when bending fully forward from the standing position. An ink grid was printed symmetrically over the midline in the L4-5 region while the individual was in the upright position, and strains were calculated from the change in dimension of the grid. The strain value was recorded at full flexion and ranged between 49 and 67 per cent.

The force in the skin corresponding to the strain at full flexion can be deduced from the previous recorded skin characteristics. As the skin appeared to strain uniformly over the width of the grid the load acting on a tab width was scaled up to act across the width of the back at the L4-5 level. The distance of the skin from the centre of rotation was estimated from measurements taken from a cadaveric specimen in the lumbar region.

RESULTS AND DISCUSSION

Farfan (1975) calculated that at flexions greater than 90°, the forward bending moment due to the head, extremities, and upper trunk about the lumbar spine was approximately 83 Nm and that the moment is balanced by the posterior ligaments and the posterior fibres of the annulus fibrosus, the paraspinal muscles being inactive (Floyd and Silver 1955). However, a third structure, the skin, can resist motion in the sagittal plane. Measurements detailed above suggest that normal skin can carry up to 5 per cent of the bending moment at the L4–5 level at full flexion.

Skin offers a massive cross-sectional area and acts at a large lever arm compared with the posterior ligaments of the lumbar spine. Thus the spinal bending moment is more sensitive to the stress developed in skin than in the ligaments.

During forward flexion the elongation of the skin can be characterized by two phases (Fig. 22.3); an initial lax phase which will readily accommodate the increased convexity of the spine during flexion, and a second phase in which the skin stiffens rapidly resisting further flexion.

During flexion the lever arm of the skin will be reduced as the bending axis centre shifts posteriorly (Rolander 1966). However, for a normal healthy motion segment, the centre of motion remains within the posterior one-third of the

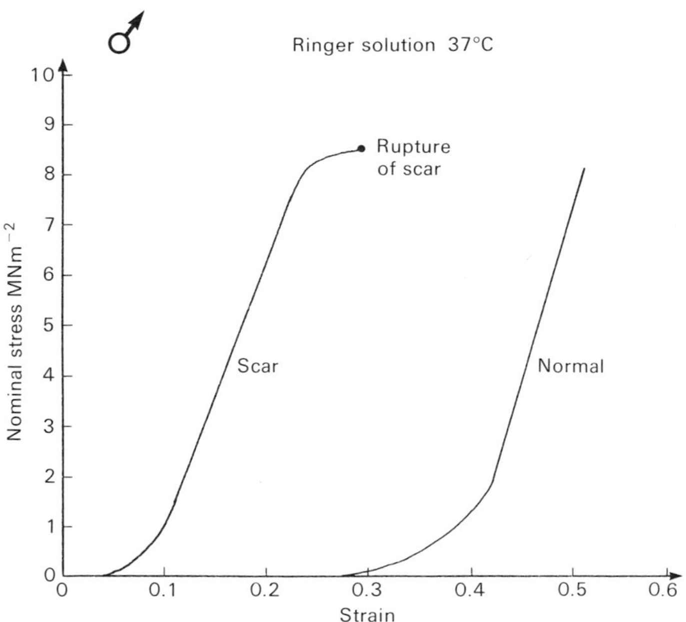

Fig. 22.3. Stress–strain relation of dermal scar tissue compared with normal adjacent skin tissue. (After Evans (1973).)

intervertebral disc and thus the reduction in leverage will be less than 10 per cent at extreme flexion.

The contribution of skin in resisting forward bending is highly dependent on the build of a person. The loose and lax skin of an obese subject would not offer significant resistance, whereas the tighter skin of a lean subject would contribute considerably more to resisting forward flexion.

As dermal scar tissue is markedly less extensible than normal skin (Fig. 22.3) it can offer a higher resistance to forward bending, which may be clinically discernible and significant.

CONCLUSION

It has been demonstrated that the muscles of the abdomen have the potential to stabilize the vertebral column through the action of the thoracolumbar fascia. This can be brought about by:

1. The intrinsic nature of the thoracolumbar fascia without direct involvement of the vertebral ligaments.
2. Muscle contraction modifying longitudinal ligament tension via the thoracolumbar fascia.
3. A combination of both mechanisms.

It has also been shown that normal skin can contribute up to 5 per cent of the bending moment developed in full spinal flexion under body weight alone.

REFERENCES

Bartelink, D. L. (1957). The role of abdominal pressure in relieving the pressure on the lumbar intervertebral discs. *J. Bone Jt Surg.* **39B**, 718–25.

Bazergui, A., Lamy, C., and Farfan, H. F. (1978). Mechanical properties of the lumbodorsal fascia. Paper No. 1A–08. Proceedings of the 1978 Society for Experimental Stress Analysis Meeting. Wichita, Kansas.

Evans, J. H. (1973). Structure and function of soft connective tissue. Ph.D. thesis, University of Strathclyde, Glasgow.

Fairbank, J. C. T. and O'Brien, J. P. (1980). The abdominal cavity and thoracolumbar fascia as stabilisers of the lumbar spine in patients with low back pain. Conference on Engineering Aspects of the Spine, Paper No. C135/80, London.

Farfan, H. F. (1973). *Mechanical disorders of the low back.* Lea and Febiger, Philadelphia.

— (1975). Muscular mechanism of the lumbar spine and the position of power and efficiency. *Orthopaed. Clins N. Am.* **6**, 135–43.

— (1977). *Mathematical model of the soft tissue mechanism of the lumbar spine. Approaches in the validation of manipulative therapy.* Charles C. Thomas, Springfield, Ill.

Floyd, W. F. and Silver, P. H. S. (1955). The function of the erectores spinae muscles in certain movements and postures in man. *J. Physiol.* **129**, 184–203.

Gibson, T., Stark, H., and Evans, J. H. (1969). Directional variation in extensibility of human skin *in vivo*. *J. Biomech.* **2**, 201–4.

Heylings, D. J. A. (1978). Supraspinous and interspinous ligaments of the human lumbar spine. *J. Anat.* **125**, 127–31.

Rolander, S. D. (1966). Motion of the lumbar spine with special reference to the stabilizing effect of posterior fusion. *Acta orthopaed. scand.* Suppl. 99.

23 'Post-Meniscectomy Inhibition': Voluntary Ischaemia Does Not Alter Quadriceps Function in Normal Subjects

M. STOKES, K. MILLS, D. SHAKESPEARE, K. SHERMAN, MICHAEL WHITTLE, AND A. YOUNG

INTRODUCTION

Quadriceps weakness following knee surgery can be due to both atrophy and reduced maximal voluntary activation (MVA). Reduced MVA, as well as producing weakness, may also contribute to atrophy. Why does this 'inhibition' of voluntary quadriceps activation occur? Pain is a possible factor (Basmajian 1970) but we have demonstrated that severe reduction of MVA can persist for several days after meniscectomy with little or no pain (Sherman *et al.* 1983) (Fig. 23.1). The presence of a knee effusion can also inhibit voluntary activation of the quadriceps (de Andrade *et al.* 1965; Stokes and Young 1984) but large effusions are only seen in a minority of patients following meniscectomy and profound inhibition is common. It has been suggested that the use of a tourniquet during surgery may contribute to post-operative weakness. The use of a pneumatic tourniquet for meniscectomy may be associated with post-operative impairment of quadriceps function and electromyographic (EMG) evidence of minor degrees of denervation (Dobner and Nitz 1982).

The aim of the present study was to determine the contribution of the effects of a tourniquet to the post-operative reduction in MVA observed in our meniscectomy patients. We therefore studied the effects of unilateral lower-limb ischaemia on quadriceps function in a small number of normal subjects.

Although our results are essentially negative, the experiment illustrates several simple approaches to the practical evaluation of quadriceps function. The techniques used examine various aspects of the muscle's physiology, viz. nerve conduction, propagation of action potential, recruitment of motor units, and force production in isolated isometric and dynamic contractions, and coordinated dynamic movement.

SUBJECTS

The experimental subjects were four of the authors (three male, one female, aged 25-36 years). None had muscle, joint, neurological or peripheral vascular

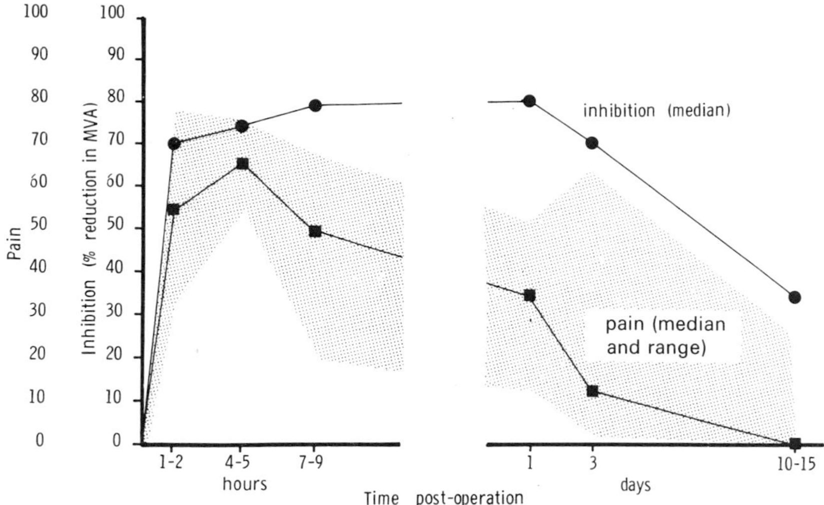

Fig. 23.1 The severity of pain and quadriceps inhibition following meniscectomy.

disease, or had sustained an injury to either leg sufficient to require immobilization of a joint for more than one week in the previous two years.

METHODS

Muscle function tests

Test 1

Needle electromyography (EMG) with a concentric needle electrode, was used to explore rectus femoris, vastus lateralis, and vastus medialis both at rest and during voluntary activity, for evidence of chronic partial denervation. Spontaneous activity (fibrillations, positive sharp waves, and bizzare high-frequency discharges) at more than two sites would have been considered abnormal. Interference pattern was assessed during a maximal voluntary isometric contraction of the muscle.

Distal motor latency was tested by stimulating the femoral nerve at the inguinal ligament and recording evoked potentials at a measured distal point in rectus femoris. Latencies were then compared between the two legs. The examiner was unaware of which leg had been ischaemic.

Test 2

The maximal voluntary activation (MVA) of quadriceps was measured, during maximal straight leg contractions, by integration of the rectified electromyogram recorded with surface electrodes at a constant site over rectus femoris (approximately mid-thigh). Skin resistance was lowered by shaving the hairs, light sand-papering of the surface to remove dead skin, and cleaning with alcohol to remove grease. The electrode sites, together with permanent skin blemishes, were recorded on transparent sheets to allow accurate relocation for subsequent testing (Dons *et al.* 1979). The subject, lying supine, was instructed to dorsiflex his foot, push his knee back towards the bed, and contract his quadriceps as hard as possible. Two two-second contractions were recorded for each leg and the maximal activity over 0.9 seconds was measured.

Test 3

Isometric quadriceps strength was measured as the force of a maximal voluntary contraction (MVC) with the subject seated and the knee at 90°. At least three contractions were made with each leg, each lasting four or five seconds. The MVC was taken to be the greatest force maintained for one second (Edwards *et al.* 1977). No visual feedback of the effort was available to the subjects.

Test 4

The force/time curve of take-off in a one-legged leap was recorded by a force plate (Kistler, Type 9821A) linked to a computer (Dec PDP-11/23). The shaded area in Fig. 23.2 was measured using a MOP planimeter (Reichert-Jung) and represents the impulse during take-off (N.s). Three leaps on each leg were performed using arm swing and knee flexion to assist take-off; the best was taken on each occasion.

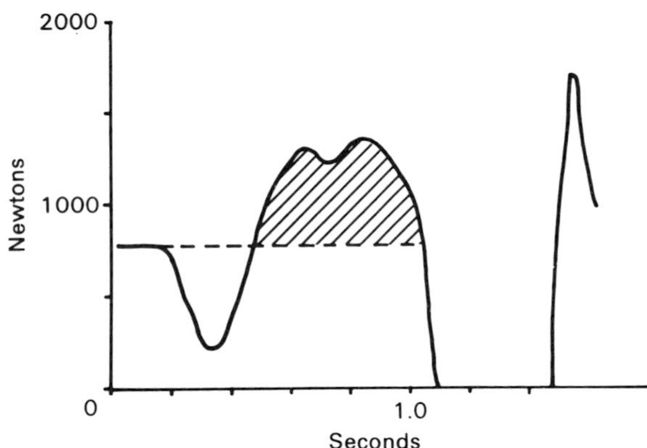

Fig. 23.2. The trace obtained from the force plate showing the force/time characteristics of a one-legged leap.

Test 5

The height of a one-legged leap was measured by standing sideways to the wall and marking the wall with an inked fingertip at full stretch before and during a vertical jump. Arm swing and knee flexion were again permitted. The best of three leaps was noted for each leg on each occasion.

Period of ischaemia

A pneumatic tourniquet was applied to the thigh by a theatre technician exactly as he would do for a patient about to undergo meniscectomy at this hospital. The subject lay supine with both knees flexed over the end of the plinth. The test leg was elevated, exanguinated with a rubber Esmarch bandage, and a pneumatic cuff (9.3 cm wide) was placed around the thigh as far proximal as possible. The leg was returned to the 'meniscectomy position' (i.e. hip in neutral and knee flexed to 90°) and the cuff was then inflated to 10 lb/sq.inch (517 mm Hg). Duration of ischaemia was for as long as each subject could tolerate, viz. 37, 38, 47, and 50 minutes.

The timing of the tests is summarized in Table 23.1.

Table 23.1. *Timing of tests*

	Leg		Test				
			1	2	3	4	5
	Test	Control	EMG	Activation	Isometric strength	Leap impulse	Leap height
Pre-ischaemia	●	■		■	■	■	■
Post-ischaemia							
Immediately	■			■			
2 hours	■			■	■		
24 hours	■	■		■	■	■	■
3 days	■	■		■	■		
3–4 weeks	■	■	■	■	■	■	■

RESULTS

None of the subjects showed any evidence of impaired quadriceps function following unilateral ischaemia. The ranges of results for MVA, MVC, and impulse and height of one-legged leaps were expressed as percentages of the pre-ischaemia values, and are shown in Table 23.2.

Table 23.2. *Results of muscle function tests*

	Range of results post-ischaemia (% of pre-ischaemia values)	
	Test leg	Control leg
Activation	94–105	94–104
Isometric strength	73–102	79–99
Leap impulse	74–118	63–112
Leap height	95–117	98–119

Needle EMG at 3-4 weeks showed no evidence of denervation. One high-frequency discharge was detected in the vastus medialis of one subject, but this was in the control leg. In another subject there was fibrillation at one site in the vastus medialis of the experimental leg, but this was not considered to be abnormal. Distal motor latencies in the ischaemic legs, expressed as a percentage of those in the control legs were, 92, 102, 108, and 110 per cent, again implying no abnormality.

DISCUSSION

The various techniques used allowed different aspects of quadriceps function to be studied. Distal motor latency (measured by femoral nerve stimulation) examined nerve conduction in the distal nerve fibres supplying the muscle. The other needle EMG test, which provided a qualitative study of activation, looked for abnornal spontaneous activity in resting muscle and also at the interference pattern. The surface EMG provided a quantitative study of voluntary activation.

Isometric quadriceps strength was measured using the simple set-up of a chair and strain gauge. A 'Cybex' machine can be used to measure both isometric and isokinetic force, but in the present study, dynamic strength was measured, in dynamic coordinated movement, as the impulse and height of one-legged leaps.

Ischaemia did not influence quadriceps MVA in our normal subjects despite the fact that the tourniquet pressure and position, and the durations of ischaemia were similar to those previously used for our meniscectomy patients (Sherman *et al.* 1983). Dobner and Nitz (1982) carried out a study in which patients underwent meniscectomy either with or without a tourniquet. The durations of ischaemia in their tourniquet group were similar to those of both our normal subjects and meniscectomy patients, but their tourniquet pressures were lower. Their patients' quadriceps were examined by needle EMG, and a progressive resistance exercise table, six weeks post-operatively. The functional tests gave significantly different results for the two groups. All of the non-tourniquet group showed normal EMG activity but 17 out of 25 of the tourniquet group did not. Our negative results in four out of four subjects were unlikely to be due to chance (Fisher's exact probability test, two-tailed, $p = 0.03$).

Electromyogram abnormalities have been detected at three weeks after meniscectomy and in some patients the EMG had returned to normal before four weeks (Saunders *et al.* 1979).

The position of the tourniquet was the same for our normal subjects and meniscectomy patients (i.e. as far proximal as possible) but neither Saunders *et al.* nor Dobner and Nitz stated the precise location of the tourniquet in their patients. Perhaps this might be an important factor if their cuffs (placed more distally) were over motor points.

An alternative explanation lies in the basic differences between normal subjects and surgical patients. The patients have received trauma to the knee and also a period of joint immobilization. Following nerve damage alternative central pathways develop and re-establishment of the former pathways is delayed if normal movement patterns are not performed soon after damage. The normal subjects were fully mobile 5–10 minutes after ischaemia whereas the patients were in bed for 2–3 days. Perhaps a combination of ischaemia, knee trauma, and immobilization is necessary to produce significant damage.

CONCLUSION

We cannot explain the absence of EMG abnormalities in our normal subjects but we can conclude that the reduced maximal voluntary activation observed in our post-meniscectomy patients is unlikely to have been caused by the tourniquet alone.

Acknowledgements

We thank Mr David Barber (Senior Operating Department Assistant, Nuffield Orthopaedic Centre) for preparing the subjects with the apparatus for ischaemia, and the Department of Health and Social Security for financial support.

REFERENCES

Basmajian, J. V. (1970). Re-education of vastus medialis: a misconception. *Archs phys. Med. Rehab.* **51**, 245–7.

de Andrade, J. R., Grant, C., and Dixon, A. St. J. (1965). Joint distension and reflex muscle inhibition in the knee. *J. Bone Jt Surg.* **47A**, 313–22.

Dobner, J. J. and Nitz, A. J. (1982). Postmeniscectomy tourniquet palsy and functional sequelae. *Am. J. Sports Med.* **10**, 211–14.

Dons, B., Bollerup, K., Bonde-Pertesen, F., and Hancke, S. (1979). The effect of weight-lifting exercise related to muscle fiber composition and muscle cross-sectional area in humans. *Eur. J. appl. Physiol.* **40**, 95–106.

Edwards, R. H. T., Young, A., Hosking, G. P., and Jones, D. (1977). Human skeletal muscle function: description of tests and normal values. *Clin. Sci. molec. Med.* **52**, 283–90.

Saunders, K. C., Louis, D. L., Weingarden, S. I., and Waylonis, G. W. (1979). Effect of tourniquet time on post-operative quadriceps function. *Clin. Orthopaed. Rel. Res.* **143**, 194–9.

Sherman, K. P., Shakespeare, D. T., Stokes, M., and Young, A. (1983). Inhibition of voluntary quadriceps activity after meniscectomy. *Clin. Sci.* **64**, 70P.

Stokes, M. and Young, A. (1984). The contribution of reflex inhibition to arthrogenous muscle weakness. *Clin. Sci.* **67**, 7–14.

24 Biplanar Radiography in Clinical Practice and in Research

M. J. PEARCY, I. PORTEK, J. E. SHEPHERD, S. J. BURROUGH, AND B. P. WORDSWORTH

INTRODUCTION

The movements of the lumbar spine are complex because of the three-dimensional structure of the intervertebral joints. Primary movements in one direction may be accompanied by movements in the other two orthogonal directions. Thus, even when the spine as a whole is voluntarily flexed, individual intervertebral joints may exhibit lateral bending and axial rotation.

To characterize fully the movements of the spine a three-dimensional technique is required using some form of stereo radiography, producing images of the spine from two separate X-ray source positions (Brown *et al.* 1976; Frymoyer *et al.*, 1979; Rab and Chao 1977; Selvik *et al.* 1976; Stokes *et al.* 1980). At Oxford a biplanar radiography system has been developed using two X-ray tubes arranged orthogonally (Pearcy and Whittle 1982).

This chapter presents the experience of using this technique over the past two years for the assessment of intervertebral movement in the lumbar spine in normals and in pathological conditions, to provide information under the following headings.

1. Basic information on intervertebral movements *in vivo*.
2. The effect of pathology on movements compared to normal.
3. Specific information about individuals of direct clinical relevance.
4. The effect of treatment for spinal disorders on the mobility of the lumbar spine.

TECHNIQUE

The technique of biplanar radiography involved a set of six radiographs taken with the subject standing in the rig shown in Fig. 24.1. Three pairs of an anterior-posterior (AP) and a lateral radiograph were taken with the subject standing: (i) upright; (ii) with his or her spine maximally flexed; (iii) with the spine maximally extended. The radiographs were analysed by marking nine bony landmarks on each vertebra in each radiograph. Two-dimensional coordinates of each landmark were obtained from both the AP and lateral radiographs using a digitizing tablet connected to a Research Machines 380Z microcomputer.

Fig. 24.1. A subject standing in the rig used for taking the biplanar radiographs.

A previous calibration of the equipment was used to compute the three-dimensional coordinates of each point. The orientation of each vertebra relative to its inferior neighbour was calculated, and comparisons of the orientations in the different positions of the subject gave the intervertebral translations and rotations that occurred as the subject moved from one position to the next.

The system was capable of measuring intervertebral translations with a root mean square (RMS) error of less than 2 mm and rotations with an RMS error of less than 1.5°. Full details of the technique and its accuracy are given elsewhere (Pearcy and Whittle 1982).

SUBJECT GROUPS

1. Normals (Pearcy *et al.* 1984)

Eleven male volunteers of mean age 29.5 years (range 25-36 years) were examined. None of them had any history of back pain requiring time off work or medical treatment, nor any episode of back pain within the preceding 12 months.

2. Low back pain and low back pain with nerve tension signs

Males between the ages of 20 and 65 years referred to the Outpatient Department of the Nuffield Orthopaedic Centre, Oxford, with a history of low back pain for at least six months were examined. Patients who had undergone lumbar spinal surgery, and those with evidence of underlying pathology (e.g. fracture, multiple myeloma, and inflammatory joint disease) were excluded.

Six patients, of mean age 35 years, with low back pain or buttock pain, without nerve tension signs were assessed.

Ten patients, of mean age 43 years, who had low back pain, sciatica, and nerve tension signs in the form of restricted straight leg raise, with or without neurological signs were examined.

3. Spondylolisthesis

Ten patients, of mean age 31.5 years (range 18–47 years), with spondylolisthesis at the lumbosacral junction (L5/S1) were investigated. The group consisted of six males and four females, all of whom had symptoms attributed to the spondylolisthesis. All the patients had bilateral defects of the pars interarticularis. The six males and two of the females had slips of grade 1, the remaining two females had slips of grade 2. (Grade 1 refers to a slip of L5 forward on S1 of 0–25 per cent, grade 4 a slip of 75–100 per cent, as described by Meyerding in 1932.)

4. Anterior fusion (Pearcy and Burrough 1982)

Eleven patients who had undergone an anterior surgical fusion procedure of one level of the lumbar spine were examined to assess the extent of bony union from 1-6 years after surgery.

5. Ankylosing spondylitis treatment trial

Fourteen male patients with ankylosing spondylitis were assessed before and after a two-week inpatient treatment programme. Treatment consisted of bed rest and exercises plus injections of corticotrophin (adrenocorticotrophic hormone—ACTH) or a placebo (normal saline), which were administered under a double-blind protocol. Within the group of 14 there were:

- Nine patients, of mean age 33 years (range 20–43 years), who received ACTH.
- Five patients, of mean age 37 years (range 26–47 years), who received the placebo.

RESULTS

1. Normal

The mean movements of the individual levels of the lumbar spine for the 11 normal subjects in flexion and extension from the upright are given in Table 24.1.

Table 24.1. *Normal subjects: mean movements of the individual levels of the lumbar spine in flexion and extension from the upright position. The translations* x, y, *and* z *are lateral shear, vertical displacement, and anterior-posterior shear respectively.*

		Translations (mm) Mean (standard deviation)			Rotations (degrees)		
	N	*x*	*y*	*z*	Flex	Lat bend	Axial rotn
Movements during flexion							
L1/2	6	0 (1)	1 (1)	3 (1)	8 (5)	1 (1)	1 (1)
L2/3	11	1 (1)	1 (1)	2 (1)	10 (2)	1 (1)	1 (1)
L3/4	11	1 (1)	0 (1)	2 (1)	12 (1)	1 (1)	1 (1)
L4/5	11	0 (1)	0 (1)	2 (1)	13 (4)	2 (1)	1 (1)
L5/S1	11	0 (1)	1 (1)	1 (1)	9 (6)	1 (1)	1 (1)
Movements during extension							
L1/2	6	1 (1)	0 (0)	1 (1)	5 (2)	0 (1)	1 (1)
L2/3	11	0 (1)	0 (1)	1 (1)	3 (2)	0 (1)	1 (1)
L3/4	11	1 (1)	0 (1)	1 (1)	1 (1)	1 (1)	0 (1)
L4/5	11	0 (1)	0 (1)	1 (1)	2 (1)	1 (1)	1 (1)
L5/S1	11	1 (1)	0 (0)	1 (1)	5 (4)	1 (1)	1 (1)

2. Low back pain and low back pain plus nerve tension signs

Tables 24.2 and 24.3 give the overall movements for the two groups and the movements of flexion plus extension for the individual levels for the two groups respectively, and comparisons between them and the normals.

Table 24.2. *Low back pain study: overall movements in degrees (mean and standard deviation—SD) of the lumbar spine in flexion and extension for the three groups; normals (Norm), low back pain alone (LBP), and low back pain plus tension signs (LBP+T), and comparisons between the groups. (NS = not significant)*

Group	Movements			Comparisons			
	N	Mean	SD		Norm v LBP	Norm v LBP+T	LBP v LBP+T
Flexion/extension							
Norm	6	67	11	*t*	2.328	6.152	2.752
LBP	6	51	13	*p*	<0.05	<0.01	<0.05
LBP+T	7	32	10				
Flexion							
Norm	6	51	8	*t*	2.317	6.596	3.393
LBP	6	38	11	*p*	<0.05	<0.01	<0.01
LBP+T	7	19	9				
Extension							
Norm	6	16	5	*t*	0.602	2.224	0.737
LBP	6	13	8	*p*	NS	<0.05	NS
LBP+T	7	11	2				

The group with low back pain alone had increased coupled movements of lateral bending and axial rotation at all levels compared to the normals. The group with nerve tension signs did not have increased coupled movements.

3. Spondylolisthesis

There were no statistically significant differences between the male and female patients and so the results for all the patients were pooled for comparison with normal subjects.

The patients had a statistically significant reduction in overall mobility compared to the normals (unpaired t test, $p<0.01$), and a reduction in flexion at all levels except L1/L2. The upper lumbar levels also showed an increase in coupled movements compared to the normals.

There was no evidence of forward or backward slip of L5 on S1 during flexion and extension.

4. Anterior fusion

Table 24.4 gives details of the patients and the results of the biplanar radiographic analysis in terms of whether bony union had occurred and of the overall mobility of the lumbar spine.

Two patients had pronounced paradoxical movements such that in voluntary flexion the fused level showed marked extension.

Table 24.3. *Low back pain study: flexion plus extension for each intervertebral level for the three groups and comparisons between the groups*

Level	Movements			Comparisons			
	N	Mean	SD		Norm v LBP	Norm v LBP+T	LBP v LBP+T
L1/2							
Norm	6	13	5	*t*	0.960	2.702	2.970
LBP	6	10	2	*p*	NS	<0.05	<0.05
LBP+T	7	7	2				
L2/3							
Norm	11	14	2	*t*	1.820	4.273	2.051
LBP	6	12	2	*p*	NS	<0.01	NS
LBP+T	9	8	3				
L3/4							
Norm	11	13	2	*t*	1.713	4.413	1.329
LBP	6	10	5	*p*	NS	<0.01	NS
LBP+T	10	7	4				
L4/5							
Norm	11	16	4	*t*	2.132	5.416	2.180
LBP	6	11	5	*p*	<0.05	<0.001	<0.05
LBP+T	10	6	4				
L5/S1							
Norm	11	14	5	*t*	2.760	4.009	0.469
LBP	6	8	4	*p*	<0.05	<0.001	NS
LBP+T	10	7	3				

5. Ankylosing spondylitis

There were no statistically significant differences between the two groups at the first assessment.

All the patients had restricted movements compared to the normals both before and after the treatment programme.

Both groups showed a trend for some movements of the lumbar spine to be increased after treatment. However, this did not reach statistical significance in the ACTH group, but was statistically significant for the group on the placebo for the movements into flexion (paired t test, $p<0.05$).

DISCUSSION

The results for the normals produced information on the physiological movements of the intervertebral joints not reported previously.

The differences between the two groups of back pain patients indicated that those with back pain alone had unilateral involvement of the muscles and

Table 24.4. *Anterior fusion: details of the patients who underwent bone grafting at either the L4-5 or L5-S1 levels and a summary of the results*

Sex	Age at operation (years)	Level of fusion	Time since operation (years and months)		Results		
					Fusion level	Lumbar mobility	Clinical assessment
F	51	L5-S1	5	2	Union	Mobile	Better
F	43	L4-5	4	2	Union	Mobile	Cured
F	39	L5-S1	2	9	Union	Mobile	Better
M	28	L5-S1	4	7	Union	Mobile	Better
F	58	L5-S1	1	4	Union	Restricted	Better
F	71	L4-5	1	3	Paradoxical	Mobile	Better
F	37	L5-S1	5	5	Paradoxical	Mobile	Cured
M	43	L5-S1	5	4	Non-union	Restricted	Same
M	50	L5-S1	0	10	Non-union	Restricted	Worse
F	48	L5-S1	5	6	Non-union	Restricted	Cured
F	27	L4-5	5	7	Non-union	Mobile	Cured

ligaments causing asymmetrical movements as shown by an increase in coupled movements. The group with back pain and nerve tension signs did not have increased coupled movements but were further restricted in flexion and extension from normal than those with back pain alone. This implied that the muscles in the group with nerve tension signs were acting symmetrically to splint the lower lumbar spine.

The patients with spondylolisthesis were assessed to answer the specific question of whether the level of the slip was mechanically unstable. Forward and backward shear of L5 on S1 was measured as a translation in the anterior-posterior direction by the biplanar analysis. In these patients with mild degrees of slip there was no evidence of forward and backward displacement during flexion and extension, and thus the term 'instability' should be reserved for the gradual slip of L5 forward on the sacrum. However, with higher grades of slip it may be possible that 'dynamic instability' does occur, but it has not been possible to investigate this as all the patients examined have had the lower grades of slip.

The patients who had undergone an anterior fusion procedure showed a trend for bony union to be accompanied by an otherwise mobile spine, and failure of union to be accompanied by restricted movements at the other levels. In this series it was also apparent that unsatisfactory clinical results only occurred with non-union. However, some of the patients were totally relieved of symptoms even though the fusion had failed to unite.

The paradoxical movements seen in two patients were probably the result of bony union occurring at the anterior margins of the vertebrae causing an alteration in the mechanics of the joint. With the vertebrae joined anteriorly the spinal muscles would be acting on a larger lever arm than normal, which would result in this level being pulled back into extension as soon as the muscles acted.

The patients with ankylosing spondylitis were assessed to examine the efficacy of the two-week inpatient treatment programme. The results showed that on the whole a slight improvement in mobility occurred in all the patients, but those receiving a placebo drug had statistically significant improvement whilst those receiving ACTH did not. This indicated that the physical therapy treatment was effective in improving mobility but questioned the efficacy of the injections of ACTH in these patients.

The results from these studies have provided information in the four categories described above. They have also outlined the clinical value of the technique in providing specific information about individuals after spinal surgery or with suspected instability, for which the technique is now used routinely. A normal pattern of movement has been defined, and for patients with spinal pathologies the technique has proved a useful research tool for evaluating the differences between groups of patients giving information on the function of the lumbar spine.

CONCLUSION

Biplanar radiography has been shown to be a useful clinical tool for the assessment of surgical fusion and mechanical stability. As a research tool it has provided information on the movements of normals and has given an insight into the effects of pathology and treatments on spinal movements and the function of the lumbar spine.

REFERENCES

Brown, R. H., Burstein, A. H., Nash, C. L., and Schock, C. C. (1976). Spinal analysis using three-dimensional radiographic technique. *J. Biomech.* **9**, 355–65.

Frymoyer, J. W., Frymoyer, W. W., Wilder, D. G., and Pope, M. H. (1979). The mechanical and kinematic analysis of the lumbar spine in normal living human subjects *in vivo*. *J. Biomech.* **12**, 165–72.

Meyerding, H. W. (1932). Spondylolisthesis. *Surg. Gynec. Obstet.* **54**, 371–7.

Pearcy, M. J. and Burrough, S. J. (1982). Assessment of bony union after interbody fusion of the lumbar spine using a biplanar radiographic technique. *J. Bone Jt Surg.* **64B**, 228–32.

—, Portek, I., and Shepherd, J. E. (1984). Three-dimensional X-ray analysis of normal movement in the lumbar spine. *Spine* **9**, 294–7.

— and Whittle, M. W. (1982). Movements of the lumbar spine measured by three-dimensional X-ray analysis. *J. biomed Engng* **4**, 107–12.

Rab, G. T. and Chao, E. Y. S. (1977). Verification of roentgenographic landmarks in the lumbar spine. *Spine* **2**, 287–93.

25 Use of ORLAU–Pedotti Diagrams in Clinical Gait Assessment

G. K. ROSE

INTRODUCTION

In 1977 Antonio Pedotti, a bioengineer, suggested that the ground reaction in FZ/FY planes from a force platform be displayed as vectors sequentially at fixed time intervals. The 'butterfly' pattern so produced (Fig. 25.1(a)) would be 'easy to read and interpret both in a quantitative and qualitative way . . . for the evaluation of normal and abnormal gait'. In addition to the force and direction of ground reaction as indicated by the vectors he made the point that:

1. Speed is indicated by the density of the vectors.
2. The pattern varies with speed in the same subject.
3. The patterns were very reproducible both in the normal and abnormal subjects.
4. Pattern recognition could possibly identify pathology.

He provided a few recordings of abnormal patterns, always starting with the diagnostic description, but did not indicate how the results could be used to modify treatment beneficially.

METHOD

It was decided to evaluate this technique in the context of the Clinical Gait Assessment laboratory at the Orthotic Research and Locomotor Assessment Unit (ORLAU) where we had available a Kistler force platform feeding into a microcomputer with a multi pen print-out. As, in appropriate cases, one of our routine procedures was to record the real-time visual vector on video (Tait and Rose 1979), it required no extra patient time or work to include this pattern in our data once the program was written.

The context of this investigation is important, where:

1. The objective of the laboratory is to assist clinical decision making; we have limited the term 'gait assessment' strictly to this process. It differs significantly from gait analysis which is the measuring of gait parameters with the manipulation and display of this information in various ways. Gait analysis is an important part of assessment which includes also history, clinical examination, the formulation of causal hypotheses, and testing the validity of these by

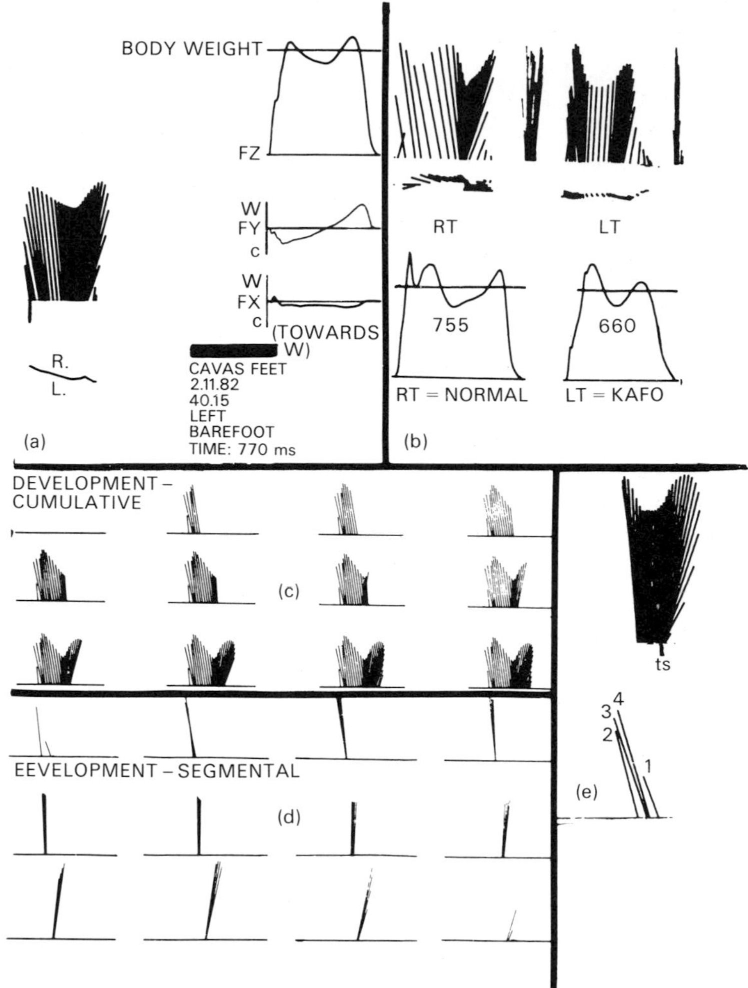

Fig. 25.1. (a) Standard ORLAU–Pedotti diagrams which includes the 'butterfly' pattern, indicator of the first vector, FZ, FY, and FX force platform print-out, diagnosis, date, video frame reference number, stance time in milliseconds and an indication of the pressure path along the foot. (b) Pedotti diagram in three ordinates. (c) Development chart—cumulative. (d) Development chart—segmental. (e) Illustrates the need to watch the print-out to obtain the sequence of vectors in primary toe strike cases.

re-examination after appropriate interference with the system. This interference includes use of orthoses (both as limitors of joint movement and postulated beneficial therapy), temporary paralysis of muscle groups, operation or physical therapy, The identification of cause from effect is particularly difficult because of the highly developed compensatory feedback which operates during gait with the pathological limitations.

2. Other available measuring techniques include: speed, step lengths, three-ordinate video with frame counter, real time visual vector, on-screen digitizer (for measurement of angles, movements, and segmental centres of mass), pedobarograph (grey and coloured), telemetry of heart rate by radio, and electromyelography by infra-red.

As we accept all comers in the orthopaedic field, the patients vary widely in age and condition, and like all who transfer measurement techniques from the normal to abnormal we soon found that problems arose. In this case:

1. A short step length, particularly combined with feet relatively close together and a poor control system, made the recording of a single-stance phase difficult using the platform longitudinally. Some help is achieved by programming as an option for transverse use.

2. Where primary toe strike occurs, as in cerebral palsy particularly where gait is slow, the overlap of vectors makes reading difficult. An optional development diagram was made available to display the 'butterfly' in 12 equal time intervals either in a cumulative (Fig. 25.1(c)) or segmental fashion (Fig. 25.1(d)). Even then the plotter sometimes had to be watched to detect the exact sequence (Fig. 25.1(e)). It was also found useful to mark the first vector routinely in all patterns.

Initally patterns were produced in all planes (Fig. 25.1(b)) but the difficulty of reading the coronal and 'overhead' views has lead us to abandon these, although we have retained the pathway vectors along the foot.

Routinely this pattern is now combined with the orthodox force platform readout (Fig. 25.1(a)). Although the 'butterfly' may look superficially similar in outline to the FZ picture, the difference is fundamental and derives from the fact that progression in the butterfly is related to distance and in the FZ to time. The difference in this relationship is very apparent in the primary toe strike print (Fig. 25.2(a)).

Results obtained in six normal subjects and 57 patients have been appraised. Patients were classified:

1. Nineteen foot conditions; deformed and/or painful.

2. Twelve arthritis of large joints. Some had failing endoprostheses. In others decisions were required in multiple joint arthritis in respect of the optional first site for replacement.

3. Twenty-six neurological conditions varying from cerebral palsy (mainly spastic) to muscular dystrophy.

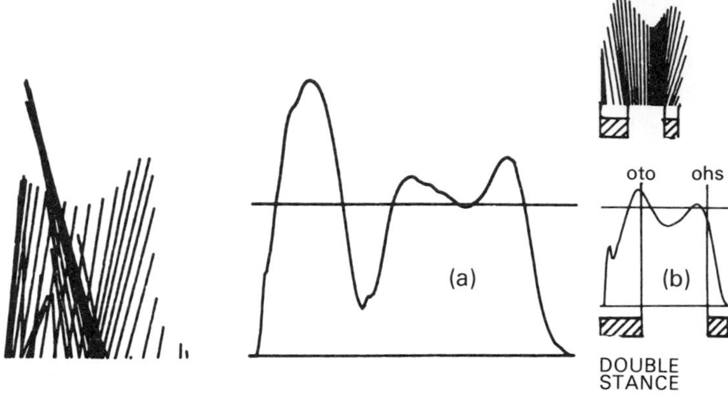

Fig. 25.2. (a) The difference between the 'butterfly' pattern and the FZ. (b) Added information—double stance (OTO = other toe-off and OSH = other heel-strike).

RESULTS

Pedotti's contention that this produces an easy to read and highly reproducible pattern was confirmed. If one already has the appropriate instrumentation, it is a simple extension without any extra burden to the patient. It is complementary to the visual vector, which has the advantage of relating the vector to other joints in space, but is difficult to appreciate as a pattern. By using the frame counter one can add to the diagram such information as double stance (Fig.25.2 (b)). It is particularly attractive to the clinician as it makes visible the real ground reaction as opposed to the abstraction of the orthodox force-platform display. It has pitfalls, of course: in particular the need to appreciate that vectors are resultants, and thus that the pathway of vectors along the foot has no direct relationship to pressures on the sole of the foot.

With regard to ease of interpretation, the bioengineer and the clinician may have different objectives and views. Clearly in mechanical terms interpretation is easy, but in clinical terms the situation is much more complicated.

The pattern of two stiff hips is clearly grossly abnormal (Fig. 25.3(a)) and it may be characteristic of this condition, but adds little to the sum total of knowledge. The most obvious use is monitoring the change of pattern with time and/or treatment. A common important question for clinicians, patients and parents is whether early deterioration is occurring; this pattern gives a simple, rapid preliminary screening most likely to be useful where change is present. Where a similar pattern is produced, this may lie within the compensatory ability of the patient and must be investigated in other ways. The clinician is particularly interested in the results of treatment and in this respect one can say whether or not the patient has come nearer to a normal pattern as illustrated by Fig. 25.3(b), a case of peroneal muscle atrophy treated initially

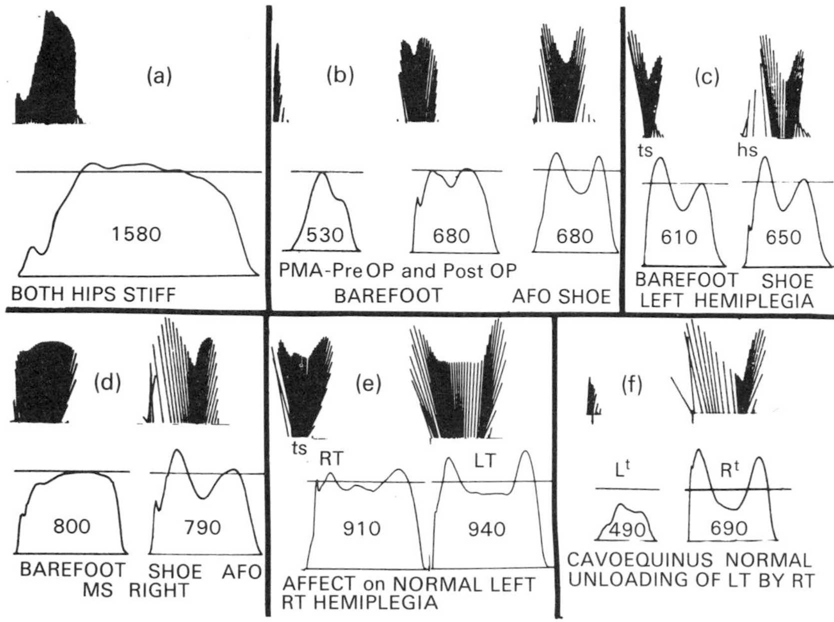

Fig. 25.3. (a) Gross disturbance of pattern with both hips stiff. (b) Case of peroneal muscular atrophy (PMA) showing progress towards normality following operative treatment and using a below-knee orthosis (AFO). (c) The difference made by using a shoe as an orthosis in a left hemiplegia. (d) Changes with shoe and AFO in multiple sclerosis. (e) The effect of a right primary toe strike from hemiplegia on the left leg pattern. (f) The unloading of the left grossly deformed and painful foot by inertial activity produced on the right.

by elongation of the tendo Achilles and subsequently by an orthosis (AFO). It can indicate the value of a shoe in hemiplegia (Fig. 25.3(c)) or that of an orthosis (AFO) plus a shoe in multiple sclerosis (Fig. 25.3(d)).

Recording of the normal limb is useful in indicating compensatory activity which may be required. In a right-sided hemiplegia (Fig. 25.3(e)) the primary toe strike produces relative lengthening of this leg. The pattern on the left looks superficially normal until the speed is noted. In such a relatively prolonged stance phase the peaks indicate firstly the high impact on landing and secondly the compensatory activity necessary to raise the body to allow the longer leg to swing through. In a patient with post-traumatic painful equino-varus of the left foot (Fig. 25.3(f)) the capability of the right to reduce the load on the left to below body weight by inertial effect is well illustrated. In a patient wearing a long leg caliper (KAFO) on the right (Fig. 25.1(b)) it will be noted that the abnormal leg can produce the more normal pattern of the two, although this is not typical.

DISCUSSION AND CONCLUSIONS

As we had anticipated, therefore, the search for the clinicians dream, akin to the philosopher's tone, of automated decision print-out was not achieved. However, we do find that the ORLAU-Pedotti diagram enlarges the wide input necessary for gait assessment, as a validating cross-reference for other data and as a factor in the complex process of formulating causal hypotheses and endeavouring to validate these. As one would expect, the nearer the pathology was to the force platform, as for example in localized foot pathology, the more direct the implication, particularly regarding the rate and rhythm of progression of vectors along the foot. Where the pathology is proximal or multiple the greater is the need to evaluate the compensatory mechanisms. Appraisal of this system has reiterated the view forced on those engaged in clinical gait assessment, namely:

1. That patterns can be recognized as characteristic of some pathological conditions, but in themselves do not add information, as the diagnosis can be made on clinical grounds. Therefore that one single form of analysis not useful.
2. That one needs a complete range of technology covering kinetics, kinematics, and energy expenditure.
3. Close collaboration between bioengineer and clinician is necessary for: (i) modification of methodology for use with the handicapped; (ii) modification of data display for maximum clinical value.

The implications of this are obvious to those contemplating the purchase of apparatus or the setting up of a laboratory. A vital factor is adequate clinical time devoted to the promotion of this subject, and this must be organized before the project starts. It is evident that some important help could be obtained by computerized clinical decision programming, probably the most important and difficult project for the future.

REFERENCES

Pedotti, A. (1977). Simple equipment used in clinical practice for evaluation of locomotion. *IEEE Trans.* **BME24**, 456–61.

Tait, J. H. and Rose, G. K. (1979). The real time video vector display of ground reaction during ambulation. *J. Med. Engng Tech.* **3**, 252–5.

26 A Tool for Standing Load Line Assessment

D. M. SMITH AND M. LORD

INTRODUCTION

For the alignment of lower-limb prostheses, the line and magnitude of the reaction force between foot and floor carries valuable information. From the product of the magnitude of the force and its distance from a particular joint axis, the external moment about that axis can be calculated; in standing, this moment must be reacted by the limb, and in walking, it contributes to the dynamics of gait. The line of the force up through the socket determines the moments applied at the socket/patient interface, and these in turn have a direct influence on socket load distribution and comfort.

Until the patient is first fitted with a new prosthesis, the limb can only be aligned by reference to positions of a number of identified points on the limb components. Reference points on the socket may be standard or transferred from features marked at the time of casting to facilitate the bench alignment. A procedure is followed to yield an estimated load distribution, according to the established practice. At the time of fitting, modifications to this alignment may be indicated in view of interaction of the patient with the prothesis. It is reasonably unlikely that the prosthetist could predict in advance the exact manner by which the patient will transfer load to a particular limb, owing to the variations in stump shape, tissue composition, habitual posture, and residual function about the remaining joints. It is during standing alignment that the prosthetist can make fine adjustments to the relative positions of the component parts to adapt for this individuality. Since in the clinic there is generally no tool available to aid the prosthetist in his assessment of the reaction force, the magnitude and line of the force must be deduced from positional observations and the subjective reports of the patient.

Various empirical rules may be cited to aid the prosthetist to describe features of the floor reaction force vector. For example, it has been suggested that a vertical line dropped from the greater trochanter will correspond to the line of the force in the sagittal plane.

Such rules may be useful practically in the clinic where no apparatus exists to quantify the reaction force vector. However, a clinically viable tool which allows measurement of these force vectors in relation to the patient would give the opportunity for a more specific description. The expected or desirable relationships between force lines and anatomical or prosthetic landmarks can be defined numerically or visualized more accurately, with a heightened awareness of the purpose and effects of the alignment adjustments. Instant feedback can be obtained on any changes in loading resulting from alignment modifications, and these observations can be taken in combination with details of the particular patient and prosthesis to build a more complete picture for the prosthetist.

The development of a clinical tool to derive this information has posed the authors certain questions relating to norms and variability of resultant floor reaction forces in relation to prosthetic alignment. How repeatable are the measures taken under clinical conditions? Can alignment changes be readily identified from features of the floor reaction? Do the estimates of load lines made by the empirical rules correspond to the actual force lines? If magnitude of load through each limb is available, what distribution between the feet might be expected? In this chapter, three experiments are described which attempt to provide such baseline information.

The motivation to answer these questions derives from the real prospect of development of such a tool within the constraints of cost and operating conditions relevant to a prosthetist's fitting room. A version of such a force-measuring device is already under development, as either a biofeedback aid for neurologically impaired patients or as a standing load line assessor for amputees (Smith *et al*. 1983).

THE DOUBLE VIDEO FORCEPLATE

The double video forceplate (DVF) is primarily a clinical tool to register the vertical component of floor reaction forces under each foot of a standing subject. It is available as a mobile, trolley-mounted system with a detachable pedestal (Fig. 26.1). In its present development form, the pedestal has two inset active areas on which the subject places his feet. Rear and medial guides are used to locate the feet in a standard orientation on these areas.

On a trolley-mounted monitor screen, two shoe outlines are drawn to correspond to the foot placement, and crosses indicate the locations of the individual centres of foot pressure (CFPs) and the overall CFP with respect to these outlines. The display is similar to that of Fig. 26.2. The percentage weight borne on each limb is indicated. These variables are derived from force transducers under the active footplate areas of the pedestal which communicate with a microcomputer linking pedestal to monitor screen and printer.

The operator can communicate with the device via a keyboard, and typically enters the patient's name, diagnosis, dominance, and shoe size in response to startup cues. The shoe size is determined by reference to markings on the pedestal footplates, and is used by the system to scale the foot outlines on the monitor screen. If the operator elects to make a recording, for the next 15 seconds the data from the pedestal are sampled, and the average CFP positions

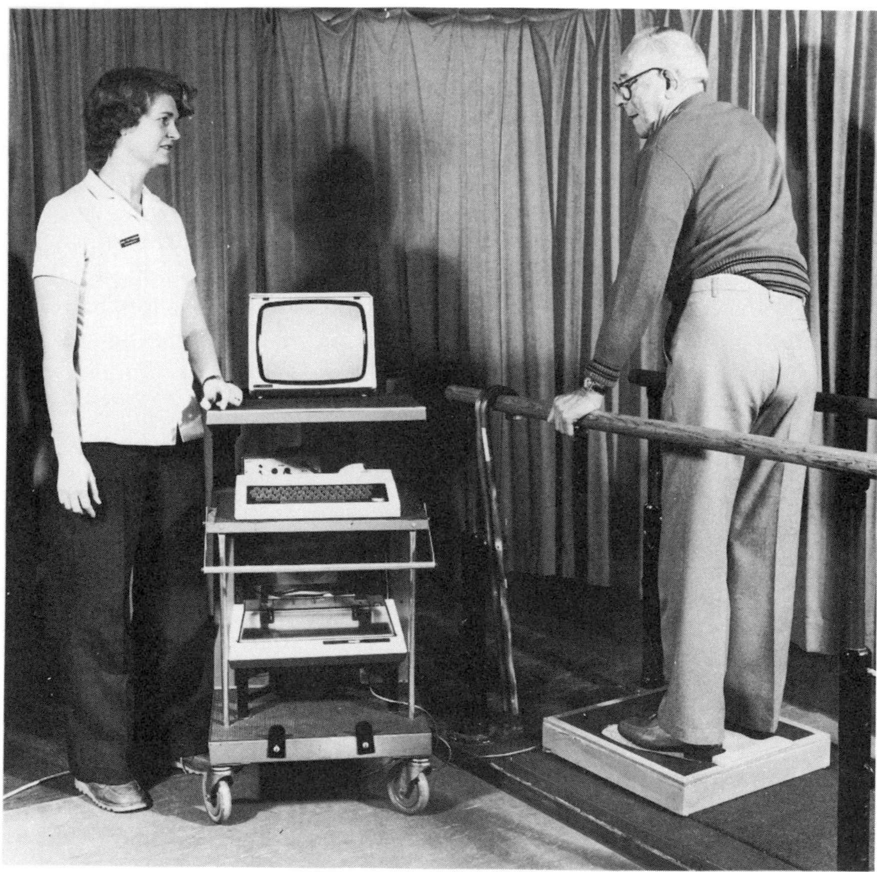

Fig. 26.1. The double video forceplate demonstrated at the Limb Fitting Centre, Roehampton. In normal use, the monitor trolley is placed directly in front of the patient.

for this period calculated. A printout of the details entered at the start, and a copy of the monitor screen image with averaged CFP positions follows automatically (Fig. 26.2).

RELATION OF THE FLOOR REACTION FORCE TO THE GREATER TROCHANTER POSITION

A series of experiments have been performed to test for correspondence, in the sagittal plane, of the line of the floor reaction force and the greater trochanter. The reason to suppose that this might exist is as follows. It is reasonable to suggest that a person would adopt a posture to minimize the need to react

NAME: MR C
HOSPITAL NO/COMMENTS:
JO RIGHT BK
DATE: 27 OCT 1982
FOOTSIZE: E
HANDEDNESS: R
66Kg
TOTAL 0m 15secs

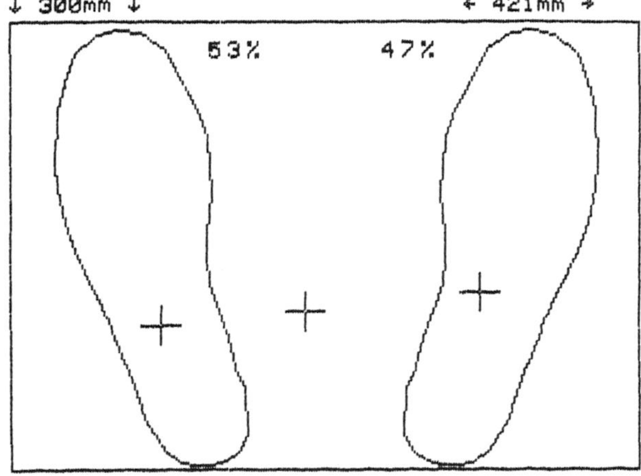

Fig. 26.2. A printout obtained from the double video forceplate. The boxed figure is similar to the monitor screen image.

external moments about the hip joints. If the force line passes centrally through the hip joint, then no external moments are generated about this joint. Furthermore the greater trochanter bears a skeletal relationship to the acetabulum, and subject to interpersonal variations in anatomy of the femur and its orientation to the pelvis, there might exist a relationship between location of the greater trochanter and line of the reaction force in order to minimize hip moments.

A comparison has been made between the anteroposterior location of the CFP and a vertical line through the greater trochanter, with reference to the rear of the heel. The use of the DVF to measure location of the CFP enforces a standard foot position during the tests, which may have influenced the results. Details of the experimental procedure are described in Appendix 26.1.

The results from 14 healthy young men are presented graphically in Fig. 26.3. Each subject was tested four times (shod, barefoot, barefoot, shod), measuring the positions of the CFP and greater trochanter for both limbs averaged over 15 seconds. The 14 × 8 coordinate pairs of positions thus generated are shown in Fig. 26.3.

The points are seen to be scattered with respect to a 45° line which represents the coincidence of the two measures. The mean values of the x and y coordinates, x = 112 mm and y = 115 mm, indicate that on average the two measures coincide, within the estimated 4 mm experimental accuracy of the measurement

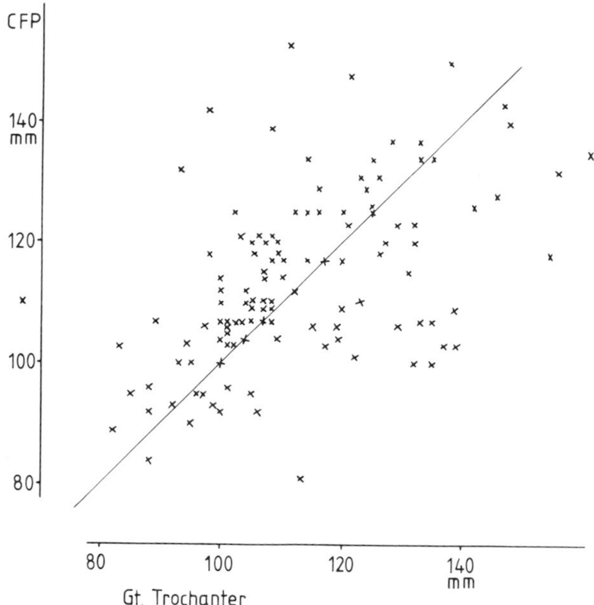

Fig. 26.3. Relationship between the horizontal location of the greater trochanter and the centre of foot pressure in the anteroposterior direction, both measured with reference to the rear of the heel. Crosses represent eight measures on 14 right-handed male subjects.

technique. However, the standard deviations of the data are large (17 mm and 14 mm) and from Fig. 26.3 it can easily be seen that the error in estimation of the load line from greater trochanter position could be as much as 40 mm in an individual case.

The data from this experiment were probed for systematic relationships, to account for the scatter about the mean values. First it was noticed visually that the four data points for the right or left foot of any individual tended either to be clustered in one area, or spread along a line. Regression lines were calculated for each set of four points. For the small clusters, a sensible line could not be drawn, and the correlations were low. Where the points were more spread and the correlations appeared more meaningful, the slope of the lines approached unity. In nine out of 28 regressions where the correlation coefficient exceeded 0.9 and the fit was significant ($p<0.05$), the slope of the line was 1.04 ± 0.13. This suggests that where a subject adopted a measurably different posture between tests, the locations of the force line and greater trochanter tend to move in unison, although they were not equal.

A regression line drawn through all the data points taken together, and then through all left foot or right foot points, showed a non-significant fit. It was not possible to separate the points statistically on the basis of right or left foot: the points were fairly evenly scattered throughout the cluster.

TEST OF THE VERTICALITY OF THE REACTION FORCE

It is normally assumed that in the sagittal plane, standing reaction forces will be to all intents and purposes vertical. If the reaction force is inclined for one limb, then the reaction force for the contralateral limb must be inclined in the opposite direction such that the net force in the anteroposterior direction is zero. In stance with neither foot in advance of the other, there is no reason to expect one foot to push forward and the other back. However, amputees frequently stand in a posture such that one foot is in advance of the other. Indeed one CFP may be in advance of the other even when the foot placement is even, as the results of the next section indicate.

In the frontal plane, the reaction forces might be expected to be inclined relative to the vertical, dependent mostly on the separation and orientation of the feet.

A set of tests were performed on 11 volunteers to determine the inclinations with the feet in free stance, constrained in the location currently used on the DVF, with one foot in advance of the other, and in a wide but comfortable stance. Details of the test procedures are given in Appendix 26.2.

In normal free standing, with the feet in the DVF positions, and in wide even stance, the inclinations in the anteroposterior direction are zero within experimental limits, as expected (mean angle for 22 observations 0.2°, 0.0°, 0.1° respectively). With one foot in advance of the other, a mean angle of 1.6° was encountered; analysis of variance confirmed the significance (left foot $F(3,27)$ = 12.6 and right foot $F(3,30)$ = 32.7: $p<0.05$) and a Tukey test indicated the advance foot position as the variant treatment. This angle of inclination is notably less than might be predicted if one assumes that the individual CFPs remain constant within the foot, and the load lines pass through a common point at hip level in the sagittal plane. For an average female with hip height 800 mm and foot length 250 mm, a half-foot length advance would result in inclination of 4.5°. In practice, the CFP probably moves forward on the rear foot, with the converse on the front foot, such that the separation of the CFPs is less than half a foot length, and further reduction in inclination would result from pelvic rotation.

In the mediolateral direction, the subjects adopted a free stance which resulted in an inclination of the force of the vertical of 1.4°. The DVF placement forced a slightly greater inclincation on some subjects with a mean inclination of 2.0°; results are significant (Wilcoxon *t*-test, t=33; $p<0.05$).

The comfortable width of 'wide' stance adopted by the subject varied considerably, with a distance of between 150 and 390 mm between the outer borders of the shoes. The inclincation in the anteroposterior direction has a mean value of 4.3° for wide stance, which is significant (right foot, $F(3,30)$ = 14.63 and left foot, $F(3,30)$ = 12.17; $p<0.05$).

OBSERVATIONS ON CENTRE OF FOOT PRESSURE LOCATION FOR AMPUTEES

The introduction of a clinical tool such as the DVF needs to be preceded by a series of experiments to determine exactly what can, and cannot, be deduced from the measurements. Norms for the behaviour of various patient/prothesis combinations can be established, not as a standard to which prostheses must be aligned but as a guide to alert the prosthetist to any unusual circumstances of loading, which may or may not be a result of deliberate action.

An opportunity arose to study a group of below-knee amputees attending the Bioengineering Centre with patellar tendon bearing prostheses and vacuum-formed polypropylene sockets (Davies and Russell 1979). Because the type of prosthesis is constant and all fittings are done by the same prosthetist, this set of patients afforded a chance to control variability. Over a three-month period 17 amputees were monitored on the DVF, all recordings being made in the same location and at the stage where the final adjustments were complete. Over the same period 61 non-amputee visitors to the Centre were recorded at the same location.

Figure 26.4 shows the average position of the CFPs, with all data normalized

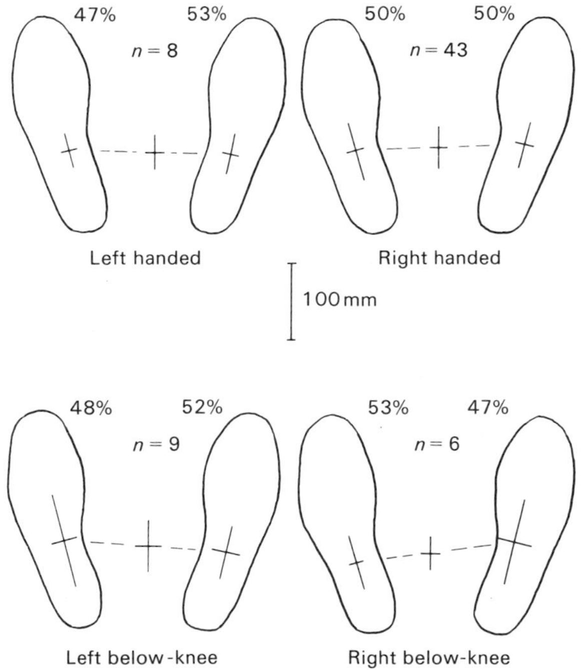

Fig. 26.4. Averaged DVF recordings on 51 normal (upper) and 15 amputee subjects with patellar tendon bearing prostheses. Length of the centre of foot pressure cross arms is ± two standard deviations.

to a standard foot length as indicated. The mean position of the overall CFP cross is fairly constant across the four groups. For the amputee population the prosthetic limb appears subject to larger variations in the individual CFP, as might be expected, with the CFP more anterior on the prosthetic side. However, for any individual the displacement of the prosthetic CFP tends to be accompanied by the opposite displacement of the sound limb CFP. Hence the overall CFP variance is not markedly different from normal.

One alignment session at delivery of a new prosthesis was observed in some detail. From a standard bench alignment, the limb was statically and dynamically aligned before it was discovered that a coupling device needed repair. The prosthesis was reassembled, bench aligned, and aligned on the patient again. The DVF printouts showed good consistency in the bench alignment and progress towards a satisfactory dynamic alignment for both sessions. The printout also showed a marked difference from the old limb which the patient had been wearing.

CONCLUSION

The observations described above are part of a continuing programme to develop the potential use of the double video forceplate in the amputee clinic, and to explore its use in other orthopaedic applications. While it is intended to continue to gather data from non-amputees and lower-limb amputees to establish norms, it is in illuminating an individual case that the DVF should provide its most useful function. Observations on its clinical use by the research prosthetist at our Centre, and by physiotherapists at the Roehampton Limb Fitting Centre are continuing.

Experience to date has indicated that reaction force information is often surprising to clinical staff, who would welcome a more fundamental understanding of its relevance and interpretation. This chapter describes a modest beginning to provide the same.

APPENDIX 26.1. METHOD OF GREATER TROCHANTER LOAD LINE EXPERIMENT

The experiment was based on the DVF. In preparation, each subject was marked on the skin over each greater trochanter with a dot approximately 5 mm in diameter. This procedure was carefully done with the subject in the standing position. The subject then stood on the DVF with hands loosely in front of the body in order not to obscure the dots. The DVF was placed so that the subject had a long-distance view out of a window, which he faced squarely, and due regard was paid to visual or attention distractions which might cause the subject to turn to one side.

From a distance of approximately 3.5 m on either side of the DVF, a narrow vertical light beam was projected against the length of the limb from marker to floor level; these beams could be deflected to move the projected vertical line forward or backward on the limb. An operator at each source kept the vertical line in registration with the dot marked over the greater trochanter during a 15-second DVF recording. Although physiological sway results in some antero-posterior movement of the dots during the test, it was possible in most cases to place the line in a mid-position of the dot from which deviations of a few millimetres would be seen.

After each recording, the position of the vertical beam with respect to the rear of the DVF heel guides was noted from its intersection with a scale on the side of the DVF. Each subject was recorded four times, twice with normal footwear and twice in stockinged feet.

At the end of the session, the DVF recordings were measured to reveal the distance of the CFP from the rear of the heel as determined from the foot outlines in the DVF print-out.

APPENDIX 26.2. REACTION FORCE INCLINATION EXPERIMENTAL METHOD

This experiment utilized a force platform based on Kistler force transducers and charge amplifiers. In each experiment, the subject was asked to stand in a specified posture with one foot on the force platform, and the other on adjacent floor with hands held loosely at the side. For the first four experiments, the left foot was placed on the platform, and for the second four experiments the right. The four experimental treatments were performed as part of a counter-balanced design. With reference to appropriate outlines marked on the platform and adjacent floor, indicating the foot positions enforced by the DVF, four foot positions were adopted, namely normal free stance, DVF stance, a wide but comfortable stance, and a stance with one foot a half-foot length in advance of the other.

Over a period of 30 seconds the vertical and two horizontal components of force were logged at 50 samples/second, and the angle of inclination were calculated and averaged for both feet in each foot position.

REFERENCES

Davies, R. M. and Russell, D. (1979). Vacuum formed themoplastic sockets for prostheses. In *Disability* (ed. R. M. Kennedi) pp. 385–90. Macmillan, London.

Smith, D. M., Lord, M., and Kinnear, E. M. L. (1983). Video aids to assessment and retraining of standing balance. In *High technology aids for the disabled* (ed. W. J. Perkins) pp. 42–55. Butterworth, London.

27 Gait Monitoring in Lower-Limb Amputees

J. C. WALL

INTRODUCTION

Lower-limb amputation has potentially profound consequences on the ability of the patient to walk. One of the major goals of the rehabilitation of such patients is the restoration of an independent form of locomotion which ideally would mean a normal walking pattern. There have been many systems designed to measure gait patterns objectively but, as Dewar and Judge (1980) point out, these have found minimal use clinically. These authors suggest that the two major problems are to decide which parameters to measure and what system to use. The information required of a gait-analysis system will vary with the nature of the pathology being studied. The temporal components of the gait cycle have proved useful in the study of lower-limb amputees (Murray *et al.* 1980; Drillis 1958; Zungia *et al.* 1972; James and Oberg 1973; Robinson *et al.* 1977; Dewar and Judge 1980 and Cheung *et al.* 1982, 1983). These temporal factors include stride trime, support time, swing time, double support time, and their derivatives. The results from the studies referred to above indicate that one of the gait characteristics of amputees is an asymmetrical walking pattern. The asymmetry can be seen in the measurements of step lengths and the various temporal factors.

Dewar and Judge (1980) developed a temporal asymmetry index as a gait quality indicator for lower-limb amputees. The technique employed was based on the clever use of the duration of both double support and swing phases to arrive at a single number which reflected the asymmetry between right and left sides. The weakness with this approach, however, is that although an asymmetry is shown to be present one cannot tell where in the gait cycle this is occurring.

Wall and Ashburn (1979) and Wall *et al.* (1981) studied the gait patterns of stroke victims and patients before and after unilateral total hip replacement. Both conditions result in asymmetrical walking patterns. As an indicator of temporal asymmetry the duration of the support phases for each side were compared. This provided a useful indication as to the extent to which the affected side was being favoured. However, it is felt that comparisons based on the various parts of the support phase might provide additional useful indicators of gait asymmetry. This chapter demonstrates the use of such indicators in monitoring the gait of lower-limb amputees.

MATERIALS AND METHODS

The subjects and system used in this study have been described in detail elsewhere (Cheung *et al.* 1982, 1983). In essence the system consists of a walkway, two sets of photoelectric beam relays, a control unit and an Apple II Plus microcomputer. The system has been designed to measure the duration of double-limb support and speed of walking.

The subjects who volunteered for this study were recent amputees undergoing a gait-training programme. These were divided into groups as follows: two unilateral above-knee amputees, four unilateral below-knee amputees, one blind below-knee amputee, and one bilateral below-knee amputee.

These subjects were tested on different occasions during the course of their gait training. On each occasion, the subject was tested on four walks after two practice walks. Results from these four walks were used to determine the average walking performance of the subject on that day. For ease of comparison, the training period has been divided into three stages. The period of training for the subjects in the study was approximately six weeks. It was thus decided to divide the training period into three two-week periods. For each subject, the results obtained within a specific period were used to determine his or her average performance for that period of time. As a group, the average performance within each specific period was computed from the corresponding subject's average performance in that period of time.

RESULTS

The results of the present study presented earlier (Cheung *et al.* 1982, 1983) showed that over the period studied all four groups improved with respect to walking speed with a concommitant decrease in stride time. The results also indicated an asymmetrical walking pattern for all the amputees studied. The degree of asymmetry was expressed as the difference in the duration of the temporal phases of gait, in terms of percentage stride time, between prosthetic and anatomic sides, or between left and right in the case of the bilateral below knee amputee. Thus, for example, support time during the initial training period for the AK group was 75.7 per cent for prosthetic side and 87.5 per cent for anatomic resulting in an asymmetry of −11.8 per cent. Although useful, it is felt that a better method of presenting these data is in the form of a ratio of anatomic to prosthetic side. Thus in the example given above the asymmetry index would be:

$$\frac{\text{Duration of support phase (anatomic)}}{\text{Duration of support phase (prosthetic)}} = \frac{87.5}{75.7} = 1.16.$$

Notice that in a truly symmetrical gait the asymmetry index calculated in this way would be 1. A number greater than 1 indicates a larger duration on the anatomic side and less than 1 is indicative that a greater length of time is spent in that phase on the prosthetic limb. The results of the asymmetry index for each of the four groups in this study are presented in graphical form in Figs. 27.1–27.4. It should be pointed out that double-braking support refers to the double-support phase immediately following heel strike.

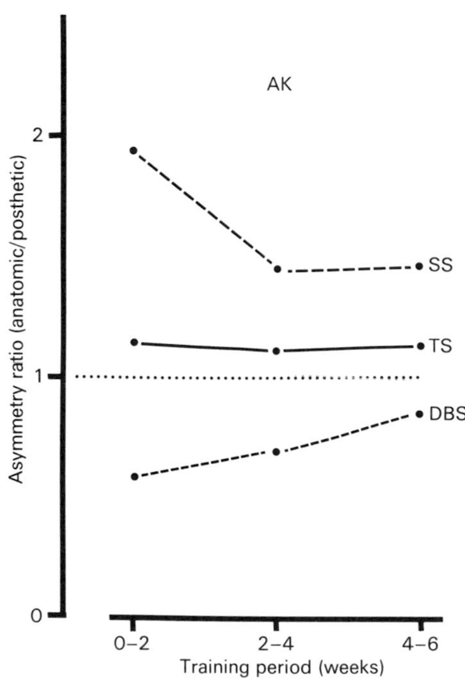

Fig. 27.1. Asymmetry ratio for two unilateral above-knee (AK) amputees. SS = single support time; TS = total support time; DBS = double braking support time.

DISCUSSION

The common pattern of asymmetry demonstrated by all four groups shows that:

1. The time spent in total support (TS) on the anatomic limb is longer than on the prosthetic limb.
2. Single-support time (SS) is also longer on the anatomic limb than on the prosthetic limb.

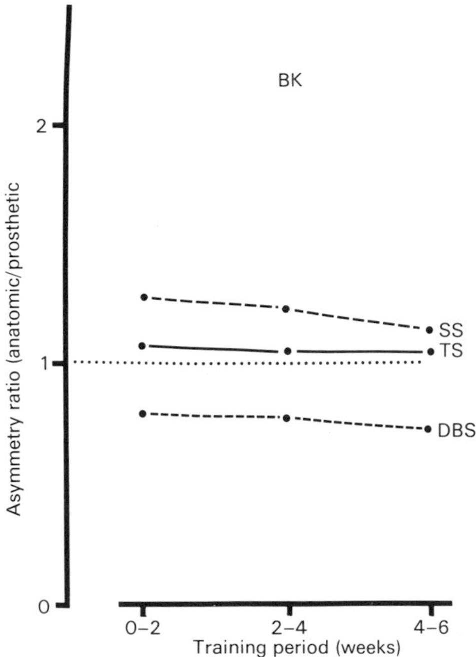

Fig. 27.2. Asymmetry ratio for four unilateral below-knee (BK) amputees.

3. The time spent in the double-braking support phase (DBS) is longer on the prosthetic limb than on the anatomic limb.

In a unilateral condition affecting the lower limbs it is common to see a favouring of one leg, resulting in a difference in stance time between right and left. This is certainly seen in the three unilateral groups in this study (Figs. 27.1-27.3). In this instance it is plausible to suggest that the reason for the shorter stance phase on the prosthetic side is due to the uncertainty of being supported on a newly acquired artificial limb. Pain in the stump during the support phase would also lead to a decreased stance time. This being the case one might expect that the part of the stance phase most likely to be affected is when the patient has to support body weight entirely on the prosthetic limb, i.e. during the single support phase. Figures 27.1-27.3 clearly demonstrate that this is indeed the case.

That the prosthetic limb has a longer double-braking support time, resulting in an asymmetry index less than 1, can be explained in terms of the need for stability. After the prosthetic limb has made contact with the ground the amputee takes time to ensure that the leg is stable prior to the single-support phase. In above-knee amputees this would mean ensuring that the knee is locked

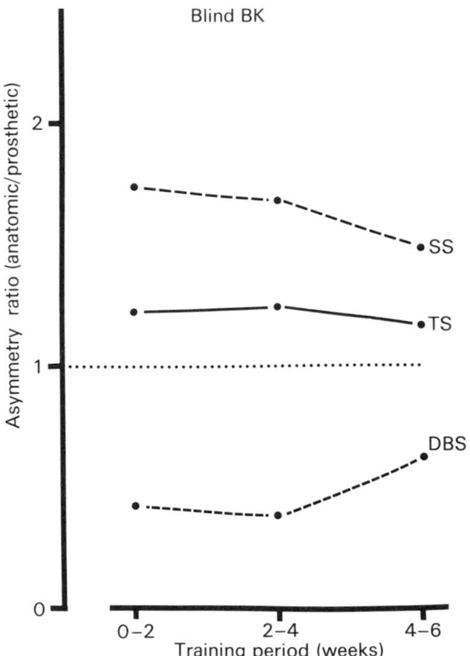

Fig. 27.3. Asymmetry ratio for one blind below-knee amputee.

before weight is transferred to it. This may help to explain why the AK group are more asymmetrical in this phase than the BK group, particularly in the first month of gait training.

It is interesting to note that the bilateral BK patient exhibits asymmetry of the same types and to a similar extent as the unilateral BK group. This patient had the left limb amputated three years prior to the right and in this case the asymmetry index was calculated as the ratio of left to right, the left side, having been fitted with a prosthesis for longer, being regarded as more 'normal'.

There are certain problems with this method of arriving at an index of asymmetry which should be mentioned.

1. They cannot be used as the sole means of monitoring walking, and consequently must be used in conjunction with other gait parameters. However, these indices do have the advantage that, when given together with the value of one of the double-support times, they allow all the temporal factors of gait to be determined.

2. The scale is non-linear: for example if double-braking support was 20 per cent on the prosthetic side and 10 per cent on the anatomic side, then the

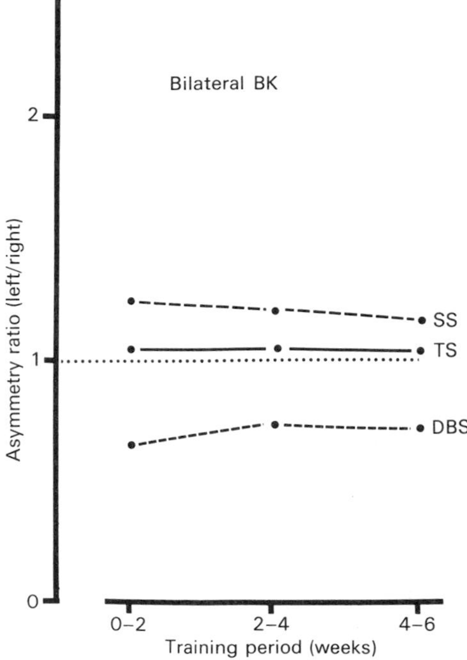

Fig. 27.4. Asymmetry ratio for one bilateral below-knee amputee.

asymmetry index calculated as suggested above would be 0.5, i.e. 10/20. However, if one were to calculate it on the basis of prosthetic/anatomic the result would be 2, i.e. 20/10. Perhaps to overcome this problem it might be better to take the reciprocal of any number less than 1 and assign it a negative value so that the scales either side of symmetry (1) would then be the same.

3. Because double-braking support time is small, typically about 10 per cent of stride time, small differences between the two legs will result in a larger asymmetry index than if the same difference occurred in the total support phase, which typically accounts for approximately 60 per cent of the gait cycle. To demonstrate this point consider a difference of 5 per cent between the two sides

	Asymmetry index
Double-braking support	15/10 = 1.5
Total support	65/60 = 1.08

Despite these difficulties it is felt that the asymmetry index proposed is a useful adjunct to other gait measurements in that the particular phases during support in which asymmetry occurs can be more readily identified. It is also considered that these indices should prove useful in monitoring progress during the gait retraining period of the rehabilitation process.

Acknowledgements

My grateful thanks are extended to Mr Cuthbert Cheung for allowing the results from his thesis to be used in this chapter. This work is continuing with the aid of a grant from The War Amputations of Canada.

REFERENCES

Cheung, C., Wall, J. C., and Zelin, S. (1982). A microcomputer based system for measuring the temporal phases of amputee gait. In *Locomotion II*, pp 20–1. Canadian Society for Biomechanics, Kingston, Ontario.

— — — (1983). A microcomputer based system for measuring temporal asymmetry in amputee gait. *Prosthet. Orthotics Int.* **7**, 131–40.

Dewar, M. E. and Judge, G. (1980). Temporal asymmetry as a gait quality indicator. *Med. Biol. Engng Comp.* **18**, 689–93.

Drillis, R. J. (1958). Objective recording and biomechanics of pathological gait. *Ann. NY Acad. Sci.* **74**, 86–109.

James, U. and Oberg, K. (1973). Prosthetic gait pattern in unilateral above-knee amputees. *Scand. J. rehab. Med.* **5**, 35-50.

Murray, M. P., Sepic, S. B., Gardner, G. M., and Mollinger, L. A. (1980). Gait patterns of above-knee amputees using constant-friction knee components. *Bull. Prosthet. Res.* **17**, 35–42.

Robinson, J. L., Smidt, G. L., and Arora, J. S. (1977). Accelerographic, temporal and distance gait factors in below-knee amputees. *Phys. Ther.* **57**, 898–904.

Wall, J. C. and Ashburn, A. (1979). Assessment of gait disability in hemiplegia. *Scand. J. rehab. Med.* **11**, 95–103.

—, —, and Klenerman, L. (1981). Gait analysis in the assessment of functional performance before and after total hip replacement. *J. biomed. Engng* **3**, 121–7.

Zungia, E. N., Leavitt, L. A., Calvert, J. C., Canzoneri, J., and Peterson, C. R. (1972). Gait patterns in above-knee amputees. *Archs phys. med. Rehab.* **53**, 373–82.

28 The Clinical use of Knee Angle and Moment Data

MICHAEL WHITTLE AND R. J. JEFFERSON

INTRODUCTION

Gait analysis has been used extensively in research laboratories for a number of decades (e.g. Bresler and Frankel 1950; Murray 1967; Simon *et al.* 1977). However, the use of cine film for kinematic studies is laborious, so that routine clinical studies have generally been restricted to simple measurements of cadence, velocity, and foot contact. The recent introduction of television-based systems has greatly reduced the time taken to analyse kinematic data, and enabled combined kinetic/kinematic studies to be performed on a routine basis for clinical assessment.

METHODS

The subject walks along a gait analysis walkway (Fig. 28.1) in the view of two television cameras. Reflective markers stuck to anatomical landmarks on the subject's legs show up as bright spots in the television image, when illuminated by a stroboscopic light which is mounted close to the camera lens and synchronized to the television linescan.

A television/computer interface determines the two-dimensional coordinates of the bright spots in the television image, and stores them on computer disc for subsequent analysis. Earlier studies were performed using an interface designed by Jarrett (1976) and a Digital Equipment Corporation (DEC) PDP-11/34 computer. Current studies are made with the commercially available 'Vicon' system (Oxford Metrics Ltd), and a DEC PDP-11/23 computer. The two cameras are used as a convergent stereoscopic pair, enabling the three-dimensional coordinates of the limb markers to be calculated to an accuracy of a few millimetres (Whittle 1982).

The walkway includes two six-component force platforms (Kistler Instruments Ltd.), which are connected to the computer through an analogue/digital converter. The ground-reaction force and the positions of the limb markers are measured every 20 milliseconds, in a common coordinate system which enables the relationships between the force vector and the joints to be defined in three dimensions.

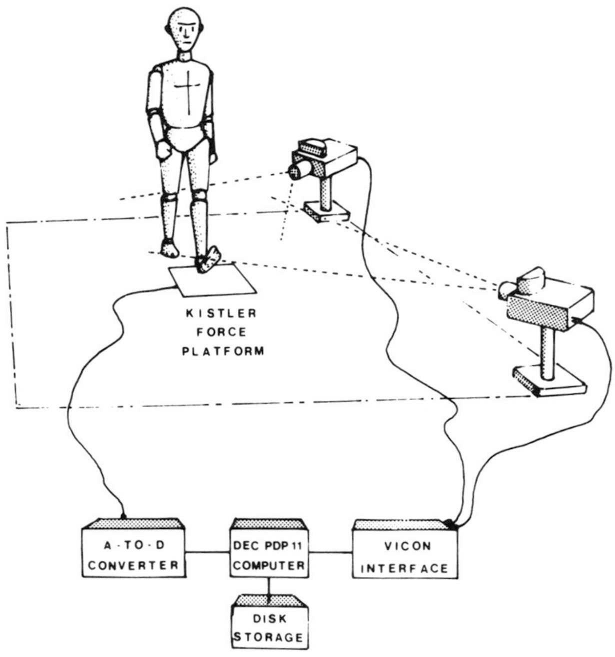

Fig. 28.1. Gait analysis system.

Although the data provided by the system could also be used to examine hip and ankle mechanics, studies to date have concentrated on the biomechanics of the knee, since it was felt that the greatest clinical value would be achieved by providing accurate information for this joint.

A display of the relative positions of the limb and the force vector in the sagittal and coronal planes (Fig. 28.2) may be useful in interpreting a pathological gait pattern, but the most useful display format to date has been the angle and moment plot (Fig. 28.3). The angle of the knee is shown in the sagittal and coronal planes, as well as the moment generated in both planes by the ground reaction force acting about the calculated geometric centre of the knee.

RESULTS

A number of research studies have been made using the system. These have been reported elsewhere (Whittle 1981), or are awaiting publication. The present chapter aims to show how knee angle and moment studies may be useful clinically.

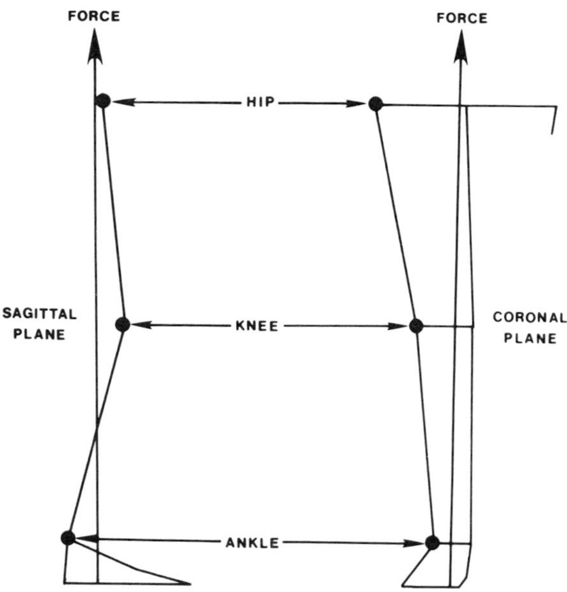

Fig. 28.2. Location of limb markers and force vector in sagittal and coronal planes.

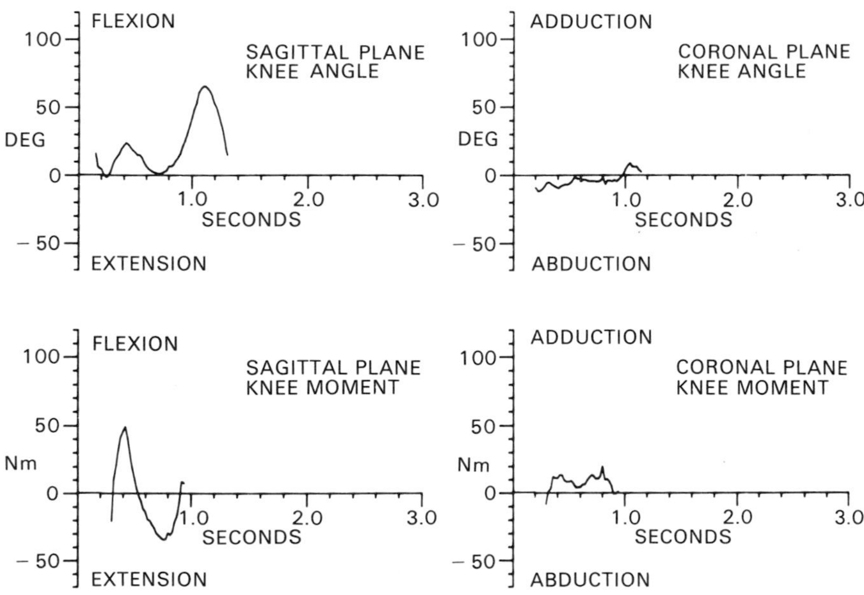

Fig. 28.3. Plot of knee angles and moments in sagittal and coronal planes—normal subject.

Case 1: Miss J.R., Age 39

This lady suffered from poliomyelitis at the age of 12 years, and has been left with extensive paralysis of the lower limbs. She walks with crutches, compensating for the lack of quadriceps power by hyperextension of the knee. Both knees have developed genu recurvatum, the right to 20° and the left to 30°. The left knee has become increasingly painful. A custom-made anti-hyperextension brace was made for her, but although limiting the hyperextension it was very uncomfortable, and caused her to walk slowly and awkwardly.

Figure 28.4 shows the sagittal plane angle and moment plots for the left knee, walking with and without the brace. They demonstrated a reduction in hyperextension during loadbearing from 36° to 20°, but also a reduction in swing phase flexion from 72° to 55°. When using the brace there was little change in the extension moment, which was very high, at about 60 Nm. The coronal plane plots were normal and are not shown.

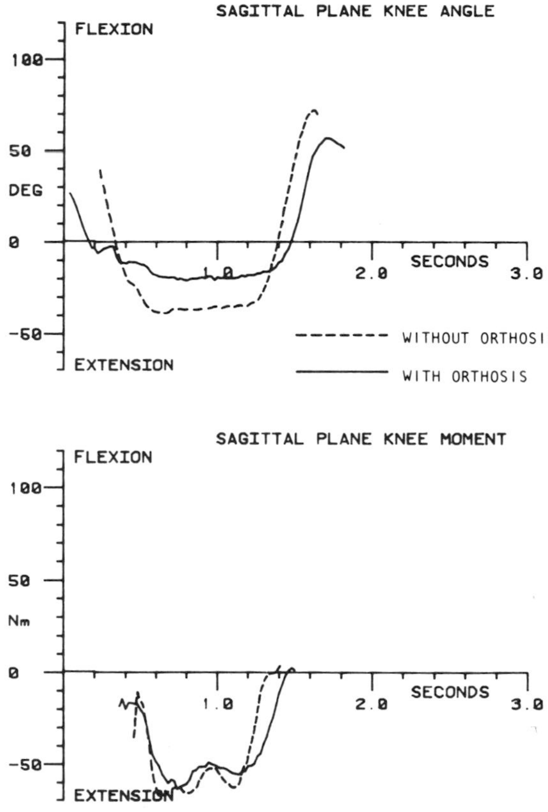

Fig. 28.4. Sagittal plane knee angle and moment with and without orthosis—case 1.

The measurements demonstrate that the orthosis has been successful in limiting the hyperextension, but the high extension moment, instead of being resisted by the capsule of the knee joint, is resisted by the orthosis, and by the soft tissues of the leg against which it bears. The loads exerted by the orthosis must be in excess of 10 kg, hence the discomfort experienced. The awkwardness and slowness in walking relate partly to this discomfort, and partly to the reduction in swing-phase flexion, causing difficulty in clearing the ground during swing-through. The reduced swing-phase flexion is probably caused by the weight of the orthosis, which as well as increasing the weight of the leg, also lowers its centre of gravity.

Case 2: Dr. S.M., Age 48

A similar case to the previous one, this lady has lower limb paralysis from poliomyelitis at the age of five years, and walks with crutches. An old fracture of the femur led to a 5° flexion contracture in the right knee, preventing her from hyperextending it during loadbearing. She has therefore adopted a crutch-walking pattern favouring the right knee. She hyperextends and takes her full weight on the left knee, which is developing genu recurvatum and becoming painful.

Figure 28.5 shows the sagittal plane knee angle and moment plots for the left knee. The coronal plane plots were normal and are not shown. The knee hyperextends to about 25° during loadbearing, and flexes to 38° during the swing phase. In contrast to the previous patient, however, the extension moment is less than 25 Nm, and it was felt that an orthosis would be able to control the hyperextension without producing excessive loads on the soft tissues. However, she probably would benefit even more from an alteration in her crutch gait to favour the left knee. This could be achieved if her right knee could be stabilized so that it could resist an extension moment.

DISCUSSION

Combined kinetic and kinematic measurement of lower limb biomechanics in three dimensions during walking has long been recognized as a powerful research tool. The advent of new technology has now made it practical to use these techniques on a routine basis in the assessment of disability. Neither kinematic nor kinetic measurements, taken in isolation, fully describe the mechanical situation. It is the relationship between the force and the position of the joint, expressed as a moment, which frequently provides the most useful information. The two examples quoted demonstrate how a knowledge of joint moments permits the interpretation of a patient's symptoms, or the prediction of the outcome of a particular type of treatment.

In order to make full use of such data, it will be necessary to establish the

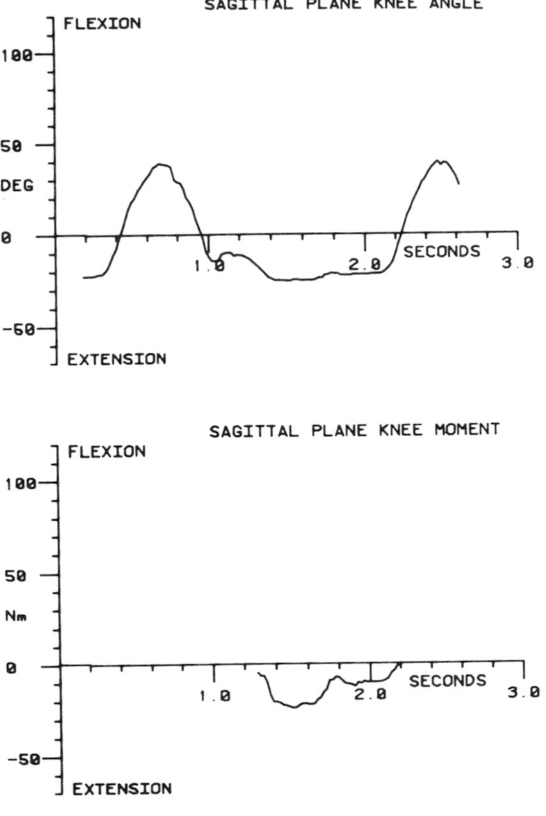

Fig. 28.5. Sagittal plane knee angle and moment—case 2.

range of normal variability for both sexes at different ages, and the types of change which are characteristic of different pathologies.

The examples given referred only to sagittal plane moments and angles, but many knee disorders involve misalignments in the varus/valgus axis. Johnson *et al.* (1980) showed the important relationship between knee alignment and compartment loading, and the need to make dynamic rather than static measurements.

CONCLUSIONS

The measurement technique is non-invasive, and usually takes about 30 minutes of the patient's time. Analysis of the results takes about another two hours. The information provided can only be obtained by the use of measuring equipment, and must be regarded as an important special investigation in certain joint disorders. It is proposed eventually to extend the system to examine both legs at the same time, and also to study the biomechanics of the hip and ankle joint.

REFERENCES

Bresler, B. and Frankel, J. P. (1950). The forces and moments in the leg during level walking. *ASME Trans.* **72**, 27–36.

Jarrett, M. O. (1976). A television/computer system for human locomotion analysis. Ph.D. thesis, University of Strathclyde, Glasgow.

Johnson, F., Leitl, S., and Waugh, W. (1980). The distribution of load across the knee—a comparison of static and dynamic measurements. *J. Bone Jt Surg.* **62B**, 346-9.

Murray, M. P. (1967). Gait as a total pattern of movement. *Am. J. phys. Med.* **46**, 290–333.

Simon, S. R., Nuzzo, R. M., and Koskinen, M. F. (1977). A comprehensive clinical system for four dimensional motion analysis. *Bull. Hosp. Joint Dis.* **38**, 41-4.

Whittle, M. W. (1981). Three-dimensional measurement of forces on the knee. In *Mechanical factors and the skeleton* (ed. I. A. F. Stokes). John Libbey, London.

— (1982). Calibration and performance of a 3-dimensional television system for kinematic analysis. *J. Biomech.* **15**, 185–96.

29 The Measurement of Normal Knee Joint Motion during Walking using Intracortical Pins

M. A. LAFORTUNE AND P. R. CAVANAGH

INTRODUCTION

In the past, most attempts to measure the motion of the knee joint have used non-invasive techniques such as exoskeletal linkages (Kettelkamp *et al.* 1970; Lamoreux, 1971; Townsend *et al.* 1977), optoelectronic systems (Antonsson and Mann 1979), roentgenography (Frankel *et al.* 1971; Soudan *et al.* 1979), and cinephotography (Eberhart *et al.* 1947; Huntington *et al.* 1979). None of these investigators were able to report patellar motion during walking but all have reported fairly similar results for movements of the tibia and femur in a sagittal plane. However, their results differ widely as far as transverse and frontal plane measurements of the motions of these bones are concerned. For instance, Huntington *et al.* computed that the knee moved from a neutral position to 25° of abduction during the swing phase of walking while Kettlekamp *et al.* reported for the same phase of gait changes from an initial 8° of adduction to the neutral position. These differences may represent the true range of variation in a normal population, they may be due to differences in definition of axis systems, or they may be a function of the measuring techniques used in data collection.

The last two possibilities are of particular interest in the present study, especially since a primary goal of the current work is to describe the motion of the patella during normal walking. Clearly this represents an extreme situation where skin markers would be of limited usefulness—since the movement of markers placed on the skin overlying the patella would bear little relation to the movements of the bone itself. Only two previous studies have reported on patellar motion relative to the femoral condyles in live human subjects (Shinno 1968; Veress *et al.* 1979) and both have involved the use of X-rays in a series of static postures of the knee joint.

There are suggestions in the literature that skin markers may also be inadequate for the measurement of some of the more intricate movements of the major bones of the knee joint. Chao (1980) stated that the relative translation between tibia and femur cannot be measured with linkages because of their physical limitations. Furthermore, he mentioned that neither skin markers nor linkages are appropriate since the relative motions of the device can lead to the recording of spurious 'apparent' joint translations. In 1977, Townsend *et al.* measured knee joint translation of almost 10 mm with a spatial goniometer, but they also reported motion of 3 mm between the linkage attachments and the bones by pre- and post-movement radiographs.

These problems were recognized as early as 1948 by Levens *et al.* who used intracortical pins to obtain more accurate information concerning the amount of axial rotation taking place in each segment of the leg during walking. They encountered several difficulties with the pin settings: these ranged from mechanical vibrations and problems with pin fixation to excessive subject pain. As a result, the data from only 12 out of 26 subjects was judged to be satisfactory. The authors confined their analysis to projections of the pins in the transverse anatomical plane and their work cannot therefore be described as being truly three dimensional. They reported that almost no transverse motion occurred between the femur and tibia after foot contact during the first 25 per cent of the cycle, and that then a slow internal rotation of the tibia with respect to the femur occurred in the second quarter of the cycle, this being followed by a rapid external rotation that terminated with toe off. During the swing phase, they reported that the tibia first internally rotated with respect to the femur and then externally rotated as heel strike approached, with the magnitude of these rotations ranging between 9° and 13°.

The purpose of the present study was to use intracortical pins to determine the three-dimensional kinematics of the tibia and patella with respect to the femur during free walking in three subjects. This chapter is part of a more extensive study that has been reported elsewhere (Lafortune 1983).

METHODOLOGY

Steinman pins 2.5 mm in diameter were inserted under local anesthetic into the corticies of the femur, patella, and tibia of the right leg in three symptom-free volunteers (Fig. 29.1). Two pins were placed in the patella and femur while a single pin was used in the tibia. Target clusters consisting of four non-colinear 10 mm diameter spheres were then attached to the pins. Preliminary details of the technique have been presented elsewhere (Lafortune *et al.* 1982). Radiographs were first used to locate the target clusters in anatomically fixed reference frames. The subjects then walked at a speed of 1.5 m/s in a previously calibrated space and were filmed by four 16 mm motor-driven cameras at 100 Hz. The resulting overdetermined system was solved for the coordinates of the targets in object space using the DLT technique (Abdel-Aziz and Karara 1971). The accuracy of the prediction of the target location was better than 0.2 mm along any of the three axes. The spatial location and orientation of the patella, tibia, and femur were determined using the procedure described by Lennox and Cuzzi (1978). The series of transformations used to obtain the relative position of the bones are shown schematically in Fig. 29.2.

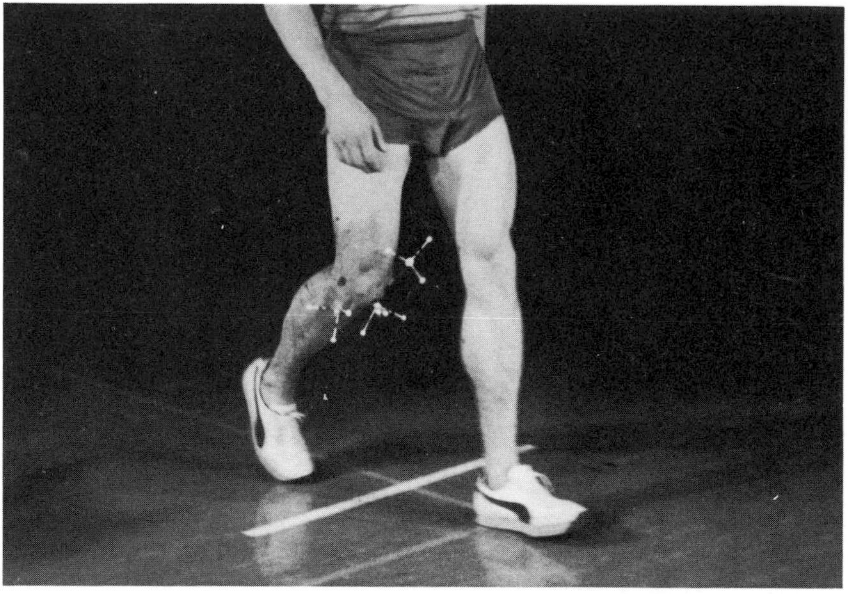

Fig. 29.1. Subject walking with the target clusters attached to the femoral, patellar, and tibial pins.

The relative motion of the tibia and patella with respect to the femur were expressed as translations and rotations around three axes (Fig. 29.3). Those axes where chosen to correspond to the joint coordinate system described by Grood and Suntay (1983). According to the conventions of that system, flexion-extension and lateral-medial translation were measured around and along the 'Xf' femoral body fixed axis; internal-external rotation and distraction were measured around and along the 'Zt' tibial body fixed axis; abduction-adduction and posterior and anterior drawer were measured around and along the floating axis 'F' which is the common perpendicular to 'Xf' and 'Zt'. The use of this joint coordinate system results in a correspondence between the measured translations and those described by clinical tests employed in the assessment of knee-joint function (Grood and Suntay 1983).

RESULTS

With one exception to be described below, the three subjects in the present study exhibited similar knee kinematics. This is emphasized by the results of all three subjects for patellar abduction-adduction shown in Fig. 29.5(b). The results presented in the remainder of the chapter are therefore those of one subject averaged over three separate trials. The relative motions of the tibia to be described will include flexion-extension, abduction-adduction and internal-

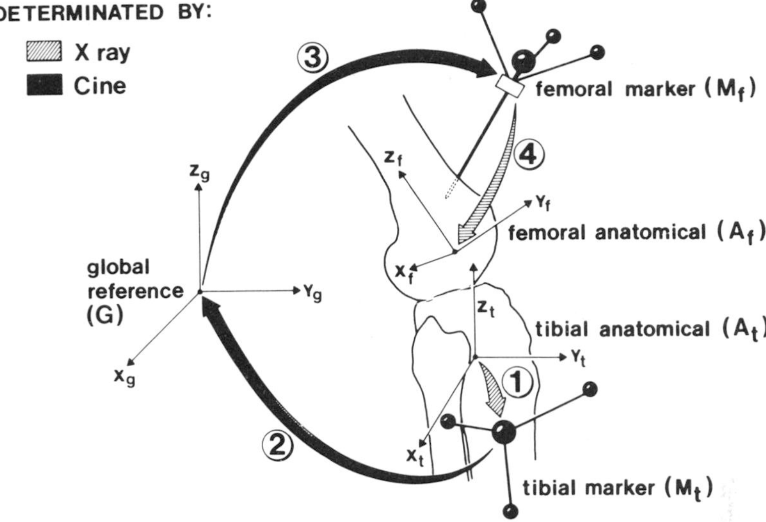

Fig. 29.2. Transformations between the various coordinate systems:
1. Express the tibial anatomical reference frame with respect to the tibial marker reference frame.
2. Express the tibial marker reference frame with respect to the global reference frame.
3. Express the global reference frame with respect to the femoral marker reference frame.
4. Express the femoral marker reference frame with respect to femoral anatomical frame.

Steps 1 and 4 are obtained from X-ray photogrammetry. Steps 2 and 3 by three-dimensional cinematography.

external rotation, while for the patella the results will be limited to shift (mediolateral translation), abduction-adduction, and internal-external rotation. These motions are considered to be the subset of the six degrees of freedom for each bone that has most clinical relevance.

Tibial motion

The knee flexion-extension which occurred around the 'Xf' femoral body fixed axis is illustrated in Fig. 29.4(a). After the heel struck the ground, the knee flexed by approximately 16° in the first 12.5 per cent of the cycle. In the following 25 per cent of cycle it extended slightly beyond the initial starting position. Thereafter flexion began—reaching 42° by toe-off (at 62 per cent of cycle time), and the flexion motion continued to a maximum value of 67°. In the last 25 per cent of the cycle, the knee extended in preparation for the next heel strike and it can be noticed that the knee had already begun to flex

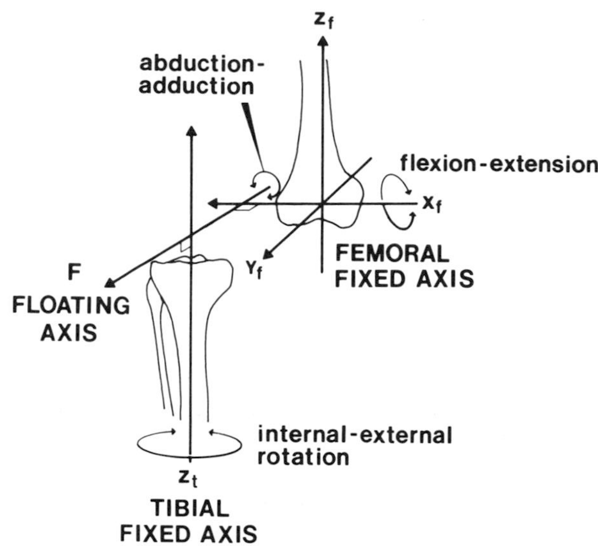

Fig. 29.3. Knee joint coordinate system proposed by Grood and Suntay (1983) 'Xf' and 'Zt' are body-fixed axes while 'F' is a floating axis.

slightly prior to touchdown. The changes in knee flexion and extension are used as a reference in Figs. 29.4 and 29.5 for subsequent discussion of other knee motions.

From heel strike to toe-off almost no abduction-adduction motion (rotation around the floating axis) took place between tibia and femur (Fig. 29.4(b)). However, after toe-off the tibia began an abductory motion that continued to increase after peak knee flexion had occurred and reached 6.5° early in the second half of the swing phase. Thereafter, the tibia adducted to reassume a neutral position at time of heel strike.

The relative rotation of the tibia (measured about the 'Zt' tibial body fixed axis) is displayed in Fig. 29.4(c). Immediately after heel strike the tibia rotated internally by 7° and it maintained that orientation until the knee reached full extension prior to toe-off. The internal rotation corresponded in time with the early knee flexion following heel strike. Next the tibia rotated internally an additional 7° reaching its peak value simultaneously with toe-off. The tibia then rotated externally until shortly before heel strike. The angular velocity of the external rotation was greater during the second part of the swing phase as knee extension was taking place. The external rotation terminated approximately 10 ms prior to heel strike.

Patellar motion

The angular displacement of the patella with respect to the femur is expressed

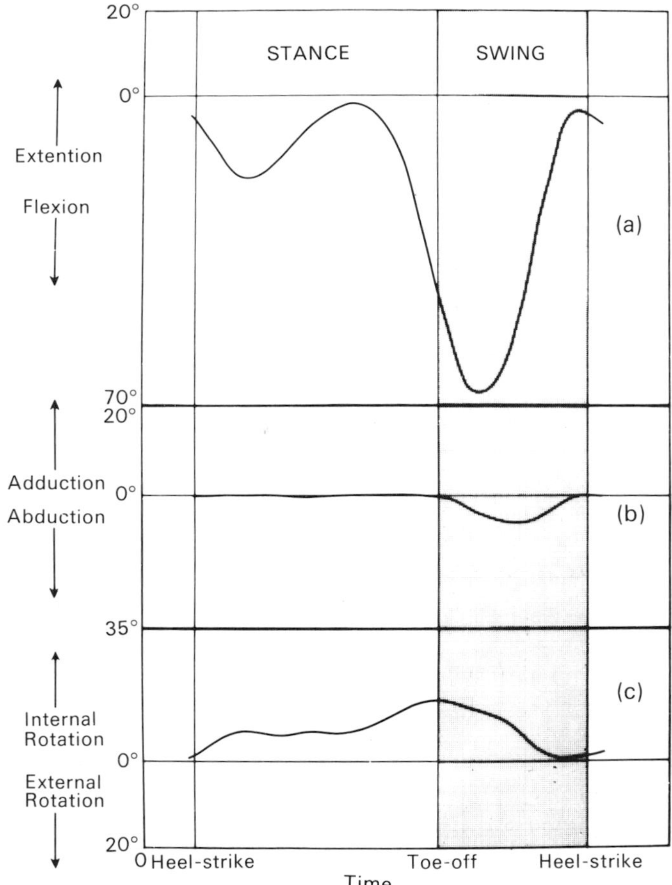

Fig. 29.4. Kinematics of the tibiofemoral joint during walking: (a) Flexion–extension. (b) Abduction–adduction. (c) Internal–external rotation.

in Fig. 29.5 relative to its position at heel strike. After heel strike the patella briefly continued its adductory motion (Fig. 29.5(b)) and for about 30 per cent of the cycle remained in a slightly adducted position. Thereafter it gradually abducted until the time of maximal knee flexion where it reached a maximum of approximately 12° of abduction.

In contrast with the other parameters presented above, the rotation or spin of the patella around its long axis showed large variability between subjects; the results from all three subjects are therefore presented in Fig. 29.5(c). Two of the subjects showed two peaks of both internal and external rotation while the other subject exhibited a gradual external rotation from heel strike until mid-swing.

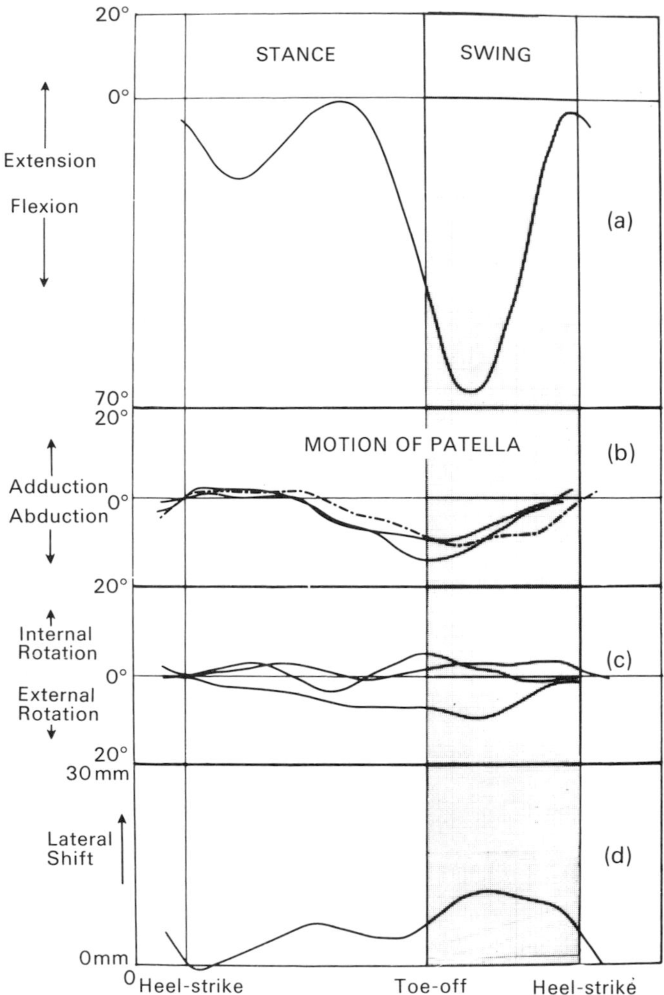

Fig. 29.5. Kinematics of the patellofemoral joint during walking: (a) Tibio-femoral flexion extension (for reference). (b) Patellar abduction–adduction. (c) Patellar internal–external rotation. (d) Patellar shift.

The translation of the patella along the 'Xf' femoral body fixed axis (shift) is expressed relative to its location at heel strike and shown in Fig. 29.5(d). At heel strike the patella started to move laterally and continued this migration for most of the first 25 per cent of the cycle. During that time, a shift of approximately 6 mm along the femoral body fixed axis was recorded. In the second quarter of the cycle, the patella experienced a slight medial translation which was followed by a major lateral shift that reached 11 mm from the initial position coincident with maximal knee flexion during the swing phase. Thereafter, the patella began a medial shift that terminated in the region of heel strike. The rate of medial shift was relatively slow in the early stage of knee extension but increased sharply as full extension approached.

DISCUSSION AND CONCLUSIONS

Of the three bones at the knee joint—the femur, the tibia, and the patella—the motion occurring between the femur and the tibia has been investigated in some detail, while the literature on the motion between the patella and femur during dynamic activities is virtually non-existent. The technique used in the present study has allowed a precise measurement of the rotational and translational behaviour of femur, tibia, and patella around and along the three orthogonal axes of a knee joint coordinate system (Grood and Suntay, personal communication).

Our results have shown that the rotational components of the motion between the femur and the tibia were consistent from trial to trial and similar from subject to subject. The pattern of flexion-extension observed among our three subjects confirms those reported by previous investigators (Eberhart *et al.* 1947; Kettelkamp *et al.* 1970; Lamoreux 1971; Huntington *et al.* 1979). However, our results on abduction-adduction and internal-external rotation disagree with some of those reported by previous investigators. Huntington *et al.* (1979) reported adductory movement of nearly 25° as the knee flexed during late support and early swing phase. These motions are completely absent in our data. Furthermore, their results on both abduction-adduction and internal-external rotation showed large trial-to-trial variations. These differences as well as the high degree of inconsistency they reported may have been caused by the use of skin markers that can move fairly independently of the underlying structure. Kettelkamp *et al.* (1970) reported consistent results from trial-to-trial. Their patterns of axial rotation were similar to our own results, but the abduction-adduction movement that they reported differed markedly from those presented here. These differences are probably a function of the exoskeletal measuring devices used by Kettelkamp *et al.* since, as mentioned by these authors, physical limitations of the devices meant that translatory components of motion could not be taken into account.

In the present study using intracortical pins, we have described the motion of the patella during the dynamic activity of walking. The components of motion of the patella presented here are shown to be rather independent of the motion of the tibia and femur; for instance, an internal rotation of the tibia does not necessarily imply an internal rotation of the patella, as exemplified by Figs. 29.4(c) and 29.5(c). Thus, our results confirm that the motion of the tibia is dependent upon structures different from those determining the motion of the patella; tibiofemoral surface contact and ligamentous restraints for the tibia, and the femoral intercondylar structure and patellar ligament for the patella (Warwick and Williams 1973).

It is interesting to note that during the knee extension phase of walking the patella moves laterally, as was reported by Reider *et al.* (1981), who performed knee extension experiments on 12 cadavers. However, there exist considerable differences between our findings and theirs when tilt (rotation of the patella around its long axis) and abduction–adduction are examined. Several factors may be responsible for the differences observed, including their choice of a set of axes fixed in the tibia, and the absence of muscular activity.

REFERENCES

Abdel-Aziz, Y. I. and Karara, H. M. (1971). Direct linear transformation from comparator coordinates into object space coordinates in close range photogrammetry. *Proc. ASP Symposium on Close-Range Photogrammetry*, pp. 1–19.

Antonsson, E. K. and Mann, R. W. (1979). Automatic 3-D gait analysis using a SELSPOT centered system. In *Advances in bioengineering*. ASME.

Chao, E. Y. S. (1980). Justification of triaxial goniometer for the measurement of joint rotation. *J. Biomech.* **13**, 989–1006.

Eberhart, H. D., Inman, V. T., Saunders, J. B., Levens, A. S., Bresler, B., and McCowan, T. D. (1947). Fundamental studies on human locomotion and other information relating to design of artificial limbs. A report to the National Research Council Committee On Artificial Limbs. University of California, Berkeley.

Frankel, V. H., Burstein, A. H., and Brooks, D. B. (1971). Biomechanics of internal derangements of the knee. *J. Bone Jt Surg.* **53A**, 945–62.

Grood, E. S. and Suntay, W. J. (1983). A joint coordinate system for the clinical description of three-dimensional motions: application to the knee. ASME *J. biomech. eng.* **105**, 136–44.

Huntington, L. J., Kendall, J. P., and Tietjens, B. R. (1979). A method of measuring from photographic records the movement of the knee joint during walking. *Engineering in medicine* **8**, 143–8.

Kettelkamp. D. B., Johnson, R. J., Schmidt, G. L., Chao, E. Y. S., and Walker, M. (1970). An electrogoniometer study of knee motion in normal gait. *J. Bone Jt Surg.* **52A**, 4.

Lafortune, M. A. (1983). A three dimensional investigation of the knee joint during walking. Ph.D. dissertation, The Pennsylvania State University.

—, Cavanagh, P. R., Kalenak, A., Skinner, S. M., and Sommer III, H. J. (1982). The use of intra-cortical pins to measure the kinematics of the knee joint. In *Human locomotion II*. Proc. of the Second Bi-annual Conference of the Can. Soc. for Biomech.

Lamoreux, L. N. (1971). Kinematic measurements in the study of human walking. *Bull. prosthet. Res.* **10**, 3–84.

Lennox, J. B. and Cuzzi, J. R. (1978). Accurately characterizing a measured change in configuration. ASME Paper No. 78–DET–50.

Levens, A. S., Inman, V. T., and Blosser, J. A. (1948). Transverse rotation of the segments of the lower extremity in locomotion. *J. Bone Jt Surg.* **30A**, 859–72.

Reider, B., Marshall, J. L., and King, B. (1981). Patellar tracking. *Clin. Orthopaed. Rel. Res.* **157**, 143–8.

Shinno, N. (1968). Statico-dynamic analysis of movement of the knee. In *Biomechanics I*. Karger, New York.

Soudan, K., Van Andekercke, R., and Martens, M. (1979). Methods, difficulties and inaccuracies in the study of human joint kinematics and pathokinetics by the instant axis concept. Example: the knee joint. *J. Biomech.* **12**, 27–33.

Townsend, M. A., Izak, M., and Jackson, R. W. (1977). Total motion knee goniometry. *J. Biomech.* **10**, 183–93.

Veress, S. A., Lippert, F. G., Hou, M. C. Y. and Takamoto, T. (1979). Patellar tracking patterns measurement by analytical X-ray photogrammetry. *J. Biomech.* **12**, 639–50.

Warwick, R. and Williams, P. L. (eds.) (1973). *Gray's anatomy*, 35th edn. Churchill Livingstone, Edinburgh.

30 Clinical and Biomechanical Assessment of Lower-Limb Forces Following Joint Replacement Surgery

T. R. M. BROWN, I. G. KELLY, J. P. PAUL, AND D. L. HAMBLEN

INTRODUCTION

Surgical treatment of a diseased joint can incur changes in both the muscle length and the lever arm with which the muscle acts about the joint. These changes will, in turn, produce changes in the muscle force required for joint functioning and hence in the force acting across the joint surface, which has been suggested as a cause for degeneration.

HIP-JOINT FORCES FOLLOWING TOTAL JOINT REPLACEMENT AND GIRDLESTONE ARTHROPLASTY

This phenomenon is well illustrated by patients who have undergone the Girdlestone (GS) procedure where, following surgical removal of the head and neck of the femur, there is a relatively large medial displacement of the hip abductors and therefore a reduction in the lever arm. Tests have been performed on nine of these patients using the Strathclyde TV computer gait-analysis system. It has been shown (Brown *et al.* 1981) that the adducting moment at the hip is reduced when compared to the contralateral side and greatly reduced when compared to the results from a group of apparently normal individuals. From Fig. 30.1 it is apparent that the force exerted by the abductor muscles, which oppose the adduction moment, is greatly reduced on the side of the arthroplasty. When a stick was not used it can be seen that there is an increase in the force exerted by the abductor musculature, but it still does not approach normal levels. The change in anatomy at the hip joint is such that an increase in abductor force would be necessary to compensate and to maintain the value of the external moment. The inability of the abductors to compensate for the anatomical changes is manifest as an abnormal gait constituted by a Trendelenburg dip, a stiff knee on the operated side, and a persistant flexion at the contralateral knee (Kelly and Brown 1982). This might be thought to require increased

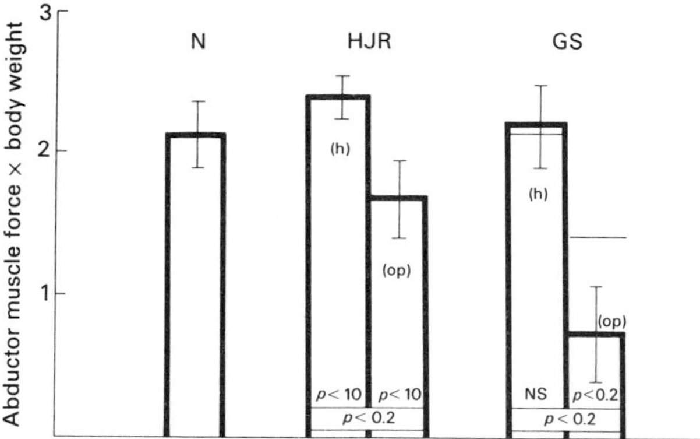

Fig. 30.1. Hip abductor muscle force at heel-stike. Maximum for normal subjects (N), hip-joint replacement patients one year post-operation (HJR), and Girdlestone patients (GS). (Mean ± 2 SE.) (h = healthy limb; op = operated limb.) (*p* values within columns re normal subjects; between columns re contralateral limb.)

quadriceps action, resulting in a supranormal value for the loading at the contralateral knee joint, but this is not the case, the reason being a reduced foot ground reaction force and hence reduced flexion moment at the knee.

Although the hip joint replacements tend to restore the functional anatomy, it has been shown that joint function does not return to normal (Brown *et al.* 1981; Murray *et al.* 1979; Stauffer *et al.* 1974). In a group of 15 patients with unilateral hip joint disease, both the ranges of movement and the forces and moments acting at the hips and knees are markedly abnormal, more than one year following arthroplasty. From Fig. 30.1 it can be seen that although there is a discrepancy between the forces exerted by the abductor musculature at the two hips, this is not as great as the difference seen in the Girdlestone group of patients. This is in keeping with the more normal anatomy seen after hip joint replacement. However, it does raise the further question as to why the abductor muscles on the side of the joint replacement arthroplasty do not function normally. Possible explanations for this are partial denervation at operation, an interruption of the propioceptive impulses from the hip joint by division of the capsules, or simply continuation of the antalgic preoperative gait. We are not in a position to confirm or refute these hypotheses. However, this impairment of abductor function results in change in the use of the lower limbs, so that there is a reduction in the knee-flexion angle in early stance in the operated limb and more flexion in the knee of the good limb throughout stance, in a manner similar to that of the GS group (Brown *et al.* 1981). Unlike the Girdlestone patients, this increased knee flexion on the contralateral side in the joint replacement patients does result in an elevation of the knee-joint forces (Fig. 30.2).

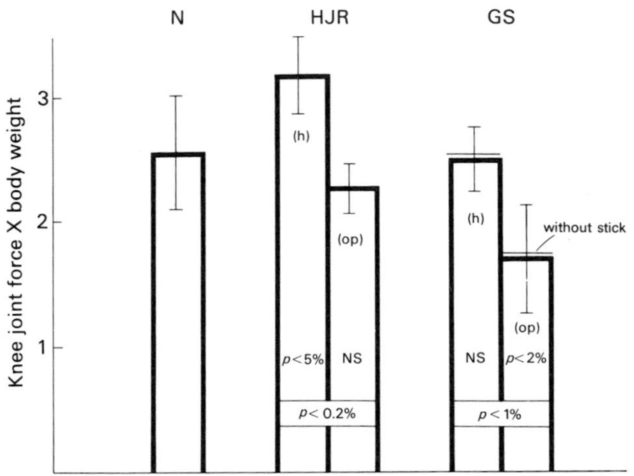

Fig. 30.2. Knee joint force at heel-strike. Maximum for normal subjects (N), hip-joint replacement patients one year post-operation (HJR), and Girdlestone patients (GS). (Mean ± 2 SE.) (h = healthy limb; op = operated limb.) (*p* values within columns re normal subjects; between columns re contralateral limb.)

EFFECT OF HIP PROCEDURES ON KNEE-JOINT FORCES

When one considers the adducting moment of the shank about the knee joint it is found that there is an increase in the adducting value of this moment in the knee of the unoperated limbs of both the hip joint replacement and Girdlestone arthroplasty groups (Brown *et al.* 1981; Kelly and Brown 1982). The increase is very much more noticeable in the knees of the Girdlestone patients when no stick was used. Since an increase in the adductor moment indicates a tendency for the load to be concentrated in the medial compartment of the knee it can be inferred that the insertion of a hip joint prosthesis may well increase the risk of developing degeneration in the contralateral knee joint and, indeed, this phenomenon has been noted clinically, especially in patients with rheumatoid arthritis. It can be further seen that a similar situation exists after excision arthroplasty of the hip, but the forces are not significantly greater than those seen in the normal control group.

The results presented here for the joint-replacement group do not distinguish between those with a Charnley or CAD Muller prosthesis. Further work is underway to evaluate the relative mechanical and physiological costs incurred by the use of these devices.

CONCLUSIONS

Observation solely of joint angles and moments do not give a full picture of total joint-loading patterns, although they may provide an indication of overall limb usage and the distribution of loading. From these studies we feel that changes in the anatomy of the hip joint results in alterations in muscle function, which in turn lead to abnormal usage of the limb. This influences the use of the contralateral limb in such a way as to alter the loading at the joints, especially the knee, and this may result in accelerated degeneration of the articular cartilage.

REFERENCES

Brown, T. R. M., Paul, J. P., Kelly, I. G., and Hamblen, D. L. (1981). Biomechanical assessment of patients treated by joint surgery. *J. biomed. Engng* **3**, 297–304.

Kelly, I. G. and Brown, T. R. M. (1982). Biomechanical assessment of patients with unilateral Girdlestone arthroplasty for failed hip arthroplasty. *J. Bone Jt Surg.* **62B**, 535–6.

Murray, M. P., Gore, D. R., Brewer, B. J., Gardner, G. M., and Sepic, S. B. (1979). A comparison of the functional performance of patients with Charnley and Muller total hip replacement. *Acta orthopaed. scand.* **50**, 563–9.

Stauffer, R. N., Smidt, G. L., and Wadsworth, J. B. (1974). Clinical and biomechanical analysis of gait following Charnley total hip replacement. *Clin. Orthopaed.* **99**, 70–7.

31 Clinical Biomechanical Assessment of the Walking Foot

M. T. MANLEY, D. ROSSMERE, AND R. DEE

INTRODUCTION

Attempts to investigate the force transmission across the foot/ground interface go back to Beely (1882), and methods used since that time include measured deformation of a walk path surface (Elftman 1934), optical techniques (Arcan and Brull 1976), and discrete pressure transducers inserted into shoes (Bauman and Brand 1963). Force-plate systems, developed to calculate the overall magnitude of foot/ground reactions, provide no information relating these forces to anatomical regions within the foot (Paul 1965). In 1974, Stokes *et al.* described a segmented force plate which attempted to monitor the segmental performance of the planted foot. However, an inked pad was required to relate the position of the foot to the recorded force data and the purely numerical output from the system was difficult to interpret. The system described below overcomes these problems and presents the data in a real-time graphic display.

METHODS AND MATERIALS

The complete system is shown schematically in Fig. 31.1. The sensing element of the system is a force plate which is made up of 24 plexiglass beams lying across the direction of walking. Each beam is supported on either end by a cantilevered load cell. These strain-gauged load cells each provide an output of 7.08 μV/N resulting in a system resolution of 0.75 N for each beam.

The force plate is fitted flush with a walkpath floor. When a subject crosses the force plate, two video cameras are used to image the plantar surface of the weight-bearing foot (Camera 1) and the lateral view of the foot and limb (Camera 2). This force data, together with both video images and four channels of EMG are correlated and displayed in real time on a CRT. The system interfaces with a computer and a video cassette recorder for both data storage and future evaluation of the data.

The system's electronics (Fig. 31.2) are controlled by an Intel 8088 microprocessor and all data manipulation is done in the digital domain. Digital processing provides high accuracy and allows direct interfacing with a gait laboratory PDP 11/34 minicomputer. The force data retrieved from the segmented force

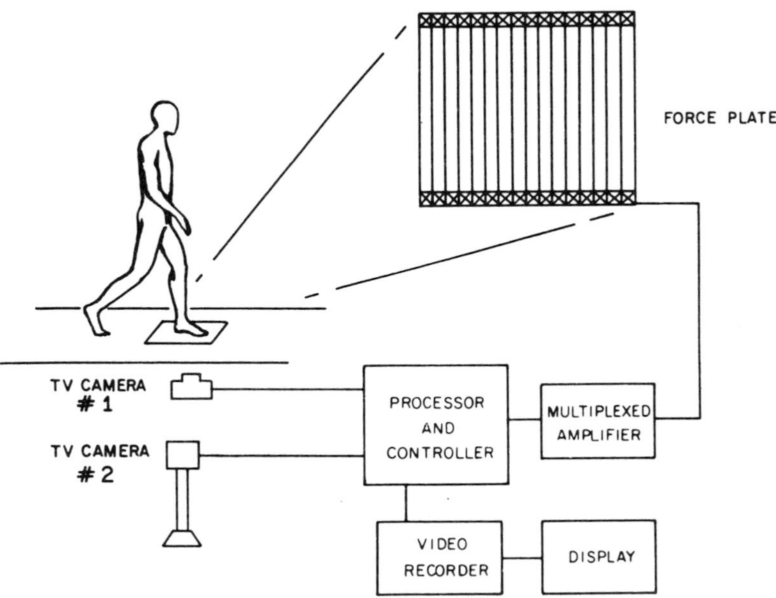

Fig. 31.1. Schematic diagram of the segmented force plate system.

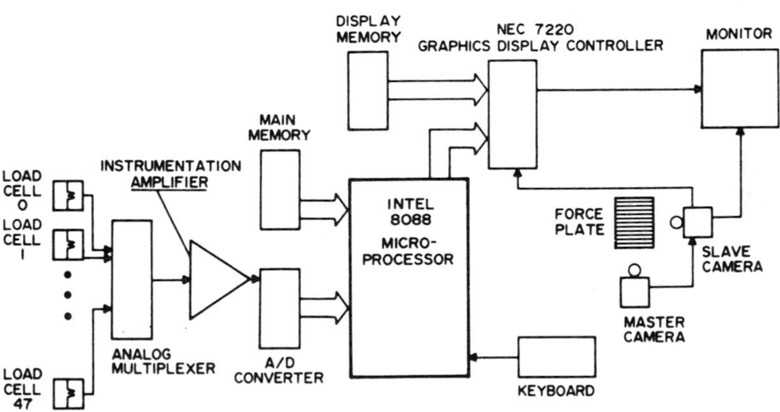

Fig. 31.2. Block diagram of the instrumentation system.

plate are first multiplexed and amplified, and then fed through a 12-bit A/D converter before being processed by the 8088. The video interface is provided by a NEC 7220 bit-mapped graphics display controller which allows the system to maintain the stringent timing requirements imposed by the correlation of real-time video images with real-time force data. Both timing and synchronization signals are generated by the 'master' video camera located underneath the force plate (Camera 1), with the 'slave' camera (Camera 2), the monitor and the VCR being synchronized to these master signals.

In order simultaneously to view two video images, the slave camera contains a screen splitter designed to break up the video/graphics display into three main sections. The 7220 graphics controller is synchronized to the master camera via external video inputs on the chip to allow graphics information to be superimposed anywhere on the display.

The display (Fig. 31.3) is refreshed every 1/60 second using 2:1 interlacing. In the top right quadrant the total force on each beam is displayed as a bar chart with each vertical bar of the chart representing the total force applied to a force plate beam.

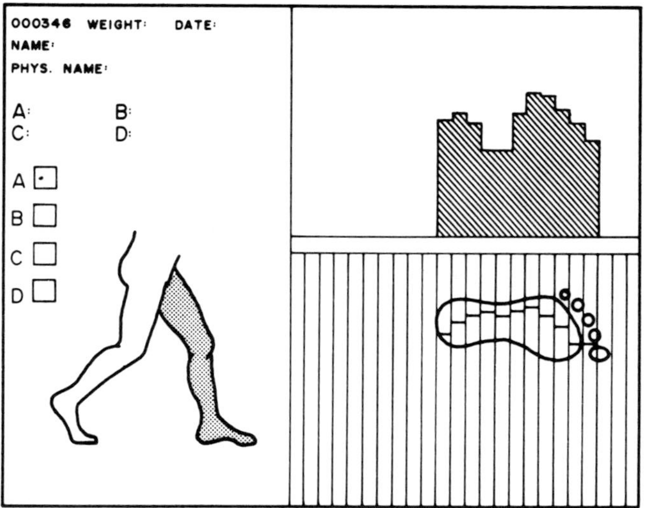

Fig. 31.3. Segmented force plate graphics display format.

In the bottom right quadrant the video image from the master camera is displayed with the centre of pressure markers superimposed onto every loaded beam of the video image. The location of these markers is calculated as $X_i = LB_i\ (A_i + B_i)$ where X_i represents the distance between the left end of the beam and the marker, L is the beam length, and A_i and B_i are the outputs from the load cells on the left and right ends respectively. The centre of pressure markers are white rectangles, four scanning lines deep, and are controlled by a clock pulse to be the width of a beam image.

The left side of the video display shows the lateral view of the walking subject along with superimposed alphanumeric data. Six lines of 30 character user-controlled alphanumerics are available to record patient parameters, while timing data is supplied by a 60 Hz frame counter.

Four channels of EMG data are displayed on the composite video image as a set of markers arranged as a column below the alphanumeric data. EMG signals are retrieved using standard surface or indwelling electrode techniques. Signals are digitized by rectifying and integrating over a 5 ms period and are used to bright-up the relevant display marker when the muscle is active.

RESULTS AND CONCLUSIONS

Once a patient has walked across the plate the recorded data can be replayed from the video cassette recorder in real time, slow motion, or frame by frame. Immediate playback allows quantitative decisions to be made about the walking foot and treatment can be prescribed or modified as necessary. In addition, the

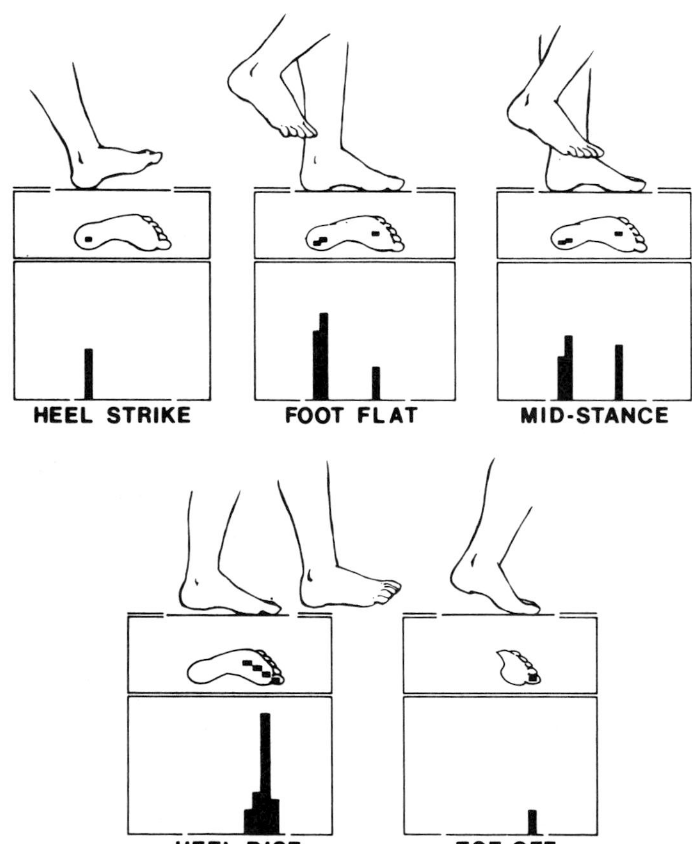

Fig. 31.4. Five instants in the stance phase of locomotion for a normal foot redrawn from recorded data. Note the progression of force along the foot and across the metatarsal heads as the stance phase progresses.

presence of four EMG markers allows immediate correlation to be made between ground-foot reactions and muscle activity, with the muscles of interest being selected by a clinician or therapist.

Figure 31.4 shows five instances in the stance phase of a normal subject redrawn from the video display. The five views show the progression of the foot through the stance phase. Longitudinal shift of the applied load from hindfoot to forefoot corresponds to the changes of shape in the bar chart. Mid-foot loading is always minimal in this normal foot. The centre of pressure markers show that normal foot loading is transferred from heel centre to the metatarsal heads and thence to the big toe. Experience with the system has shown that this normal loading pattern can be minimally or grossly disturbed by foot abnormality, with areas loaded out of sequence, or with bony prominences loaded excessively. For example, a very high magnitude of forefoot loading been demonstrated at the site of plantar ulceration in diabetic patients.

The greatest advantage of this system is that conclusions can be made about the performance of the walking foot while the patient is still in the clinic. Thus the biomechanical effects of orthotic devices such as shoe inserts, braces, and calipers can be demonstrated as soon as fitting has been completed, and corrections or modifications made to these aids as necessary. The system can also be used as a scanning device for full gait analysis assessment and is helpful in reducing the workload in our gait laboratory.

Acknowledgements

The authors wish to express their thanks to the Max and Victoria Dreyfus Foundation and the New York State Science and Technology Foundation for funding the design and construction of the segmented force plate system described in this chapter.

REFERENCES

Arcan, M. and Brull, M. A. (1976). A fundamental characteristic of the human body and foot—the foot ground pressure pattern. *J. Biomech.* **9**, 453.

Bauman, J. H. and Brand, P. W. (1963). Measurement of pressure between foot and shoe. *Lancet* **i**, 629.

Beely, F. (1882). Zur Mechanik des Stehers. *Arch. Klin. Chir.* **27**, 457.

Elftman, H. (1934). A cinematic study of the distribution of pressure in the human foot. *Anat. Rec.* **59**, 481.

Paul, J. P. (1965). Bioengineering studies of the forces transmitted by joints. Ph.D. thesis, Bioengineering Unit, University of Strathclyde, Glasgow.

Stokes, I. A. F., Hutton, W. C., and Evans, M. (1974). Force distribution under the foot—a dynamic measuring system. *Biomed. Engng* **9**, 140–3.

32 The Use of the Vector Stereograph in the Three-Dimensional Measurement of the Lower Limb

J. C. T. FAIRBANK, P. B. PYNSENT, J. ALBERT, AND H. PHILLIPS

INTRODUCTION

The measurement of the human frame in three dimensions is important for the assessment of both normal and pathological morphology. This chapter describes the use of the vector stereograph applied to the measurement of the lower limb in studies of adolescent knee pain and in the correction of lower-limb deformity following knee arthroplasty.

In clinical practice, orthopaedic surgeons have normally relied on goniometers and measurements made from radiographs in the assessment of lower-limb deformity. These methods are notoriously inaccurate. In addition, radiographs carry distortions inherent in a point source of X-rays projecting on to a two-dimensional plate, and even in skilled hands it is possible to make serious mistakes in the positioning of implants and in corrective osteotomies.

There are a number of methods of defining three-dimensional structures (e.g. Herron 1972; Takasaki 1975; Clark 1976; Fuchs *et al.* 1977; Lindström *et al.* 1982). In current studies we elected to use a method of three wires attached to a wand, with potentiometers as wire-length transducers (Clark 1976). This device was called a vector stereograph by Morris and Harris (1976), who applied it to the measurement of spine shape and movement (Grew and Harris 1979). We adapted the vector stereograph so that it could be used with a microcomputer (BBC) for the measurement of anatomical points in the lower limb.

METHOD

The overall arrangement of the equipment has been described by Pynsent *et al.* (1983). Briefly, this consists of a vertical triangular frame with a 10-turn ceramic potentiometer at each corner, directly connected to bobbins. On these bobbins are reeled nylon threads, and tension is maintained with constant-torque springs. The free ends of the three nylon threads run to the tip of a 'wand' or pointer. The wand can be moved through three dimensions, giving the basic transducer.

When the pointer is in the correct position, signal acquisition is initiated by a foot switch. There is a cancel button built into the frame, and the software allowed sequential cancelling back through the acquired points. A display monitor gave prompts for each point to be measured; additionally each time the foot switch or cancel button was used the computer gave an audible response. This system allowed the sequence of 26 anatomical landmarks on an individual to be recorded in less than one minute.

On completion of the sequence of measurements, the Cartesian coordinates of each anatomical landmark from each leg were compared, and if a greater than 5 per cent difference was recorded, a warning alarm sounded. The measurements could then be repeated if necessary. This helped to reduce errors due to patient movement during the measurement period.

For the knee measurements, the subject stood in an adjustable jig, with the pelvis resting forwards on a bar, holding it square, and with the medial borders of the feet held parallel. The knees were held fully extended and a steadying strap was used when necessary.

Having defined the geometry of the vector stereograph, the calculation is performed in three stages. First, a routine for translating the raw digital data into Cartesian coordinates; second, a procedure for the measurement of the distance between two points defined in Cartesian space, which is simply Pythagoras' theorem in three dimensions; and thirdly, an algorithm calculates the angles between three selected datum points.

APPLICATIONS

Adolescent knee pain

Anterior knee pain arising in the patellofemoral joint is common in adolescents and young adults. Its aetiology is unknown, and the symptoms relate poorly to pathological changes in the patellofemoral joint (chondromalacia patellae) (Leslie and Bentley 1978). Many authors have suggested that the symptoms may be due to abnormalities in the mechancis of the patellofemoral joint, for example, an increase in the normal valgus angularity at the knee joint, or an increase in the quadriceps, or 'Q-angle' or an increase in femoral neck anteversion. We used our vector stereograph to make accurate measurements from 14 subcutaneous bony points in each lower limb of 446 adolescents standing in the jig described above. One hundred and thirty-six had knee pain in the previous year, and 40 had complained to their doctor of their symptoms.

We were unable to demonstrate any significant differences in the morphology of the lower limbs of those with and those without a history of knee pain, even in those who had sought medical advice for their symptoms. There were no significant differences in the measurements of joint mobility, height or weight between the two groups. There was a significant difference in 'sport enjoyment' between the two groups, and it was clear that adolescents with knee pain tended to be much more active in sport than those without.

We were able to calculate that sporting activity rather than any inherent deformity was the dominant factor in the genesis of knee pain in this age group. Full details of this study are published elsewhere (Fairbank *et al.* 1984).

Measurement of lower limb deformity before and after Freeman–Samuelson surface knee arthroplasty

Lower-limb deformity has been traditionally assessed by the following three methods:

1. Clinical and goniometric measurement.
2. Weight-bearing radiographs of the knee joint.
3. Specialized long-leg radiographs.

These techniques have the disadvantage of inaccuracy (1); radiographic distortions (2,3); expense (3); special equipment requirements (3); or problems in the assessment of deformity in the presence of fixed flexion deformity of the knee and hip joints (1,2,3). We used the vector stereograph to overcome some of these problems.

The patients stood in the vector stereograph, and as far as possible the feet were placed parallel with one another. The measurement points were the anterior superior iliac spine, the greater trochanter, the medial and lateral borders of the knee joint, and the medial and lateral malleoli.

The perfectly aligned limb was taken to have a straight line passing from the hip joint, through the mid-point of the knee to the mid-point of the ankle (Kapandji 1974). Freeman takes the hip joint as lying 3 cm medial to the anterior superior spine (operative instructions). Deviations of the centre of the knee from a line joining the hip joint to the centre of the ankle joint may be expressed as an angular (Fig. 32.1) or a linear deviation (Fig. 32.2). These deviations were measured pre- and six months post-operatively.

Nine patients were studied, eight with unilateral and one with bilateral knee replacements. Eight of the knees replaced were deformed by rheumatoid arthritis and two by osteoarthrosis. Seven knees were not operated on within the study period, and served as controls. In one remaining knee, which had not been operated on during the period of the study, the joint destruction was too severe for valid measurements to be made.

The correction of deformity by the operation is shown in Figs. 32.1 and 32.2. Eight of the 10 surgically treated knees were corrected to less than 5° angular deviation or 20 mm linear deviation from the datum. Two patients had unacceptable correction. One had had a previous double osteotomy, and had a severe preoperative varus deformity of the distal femur. The other had a

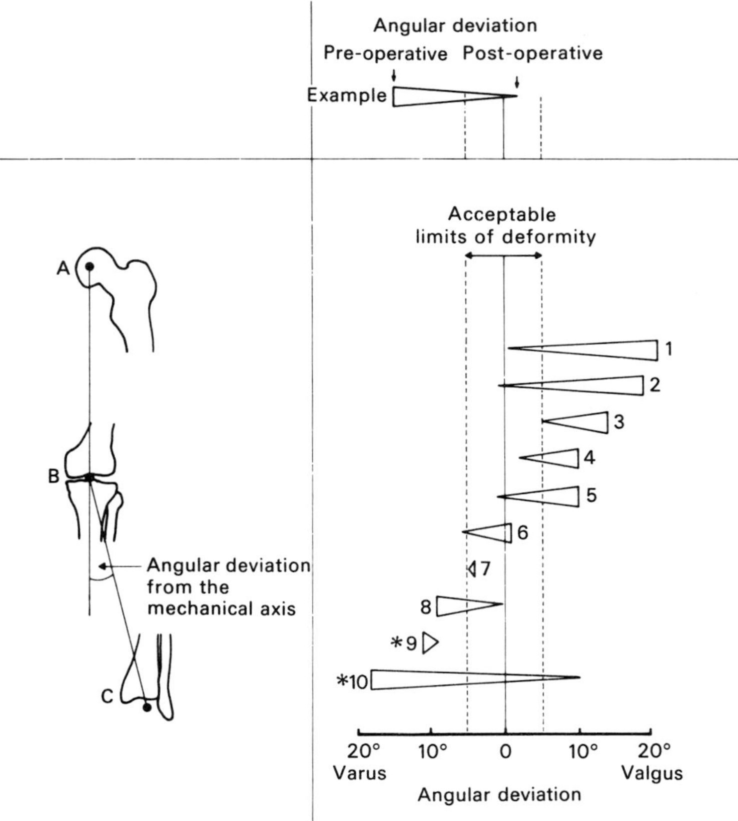

Fig. 32.1. Angular deviation of the knee: method of calculation and pre- and postoperative results from nine knees.

40° fixed flexion deformity of the knee preoperatively, and 20° deformity post-operatively rendering accurate measurement difficult.

The repeatability of the control measurements made at an interval of six months is shown in Figs. 32.3 and 32.4. No allowance is made for any progression in deformity between the measurements.

DISCUSSION

We wish to concentrate on the value of the vector stereograph in assessing the lower limb in orthopaedic practice, and not to discuss the results of our clinical studies in detail, except to illustrate the advantages and disadvantages of this system. We have found the vector stereograph to be of value in making rapid and accurate three-dimensional measurements in adolescents and young adults. Our system was accurate in measuring 1 mm differences over 1 m (0.1 per cent).

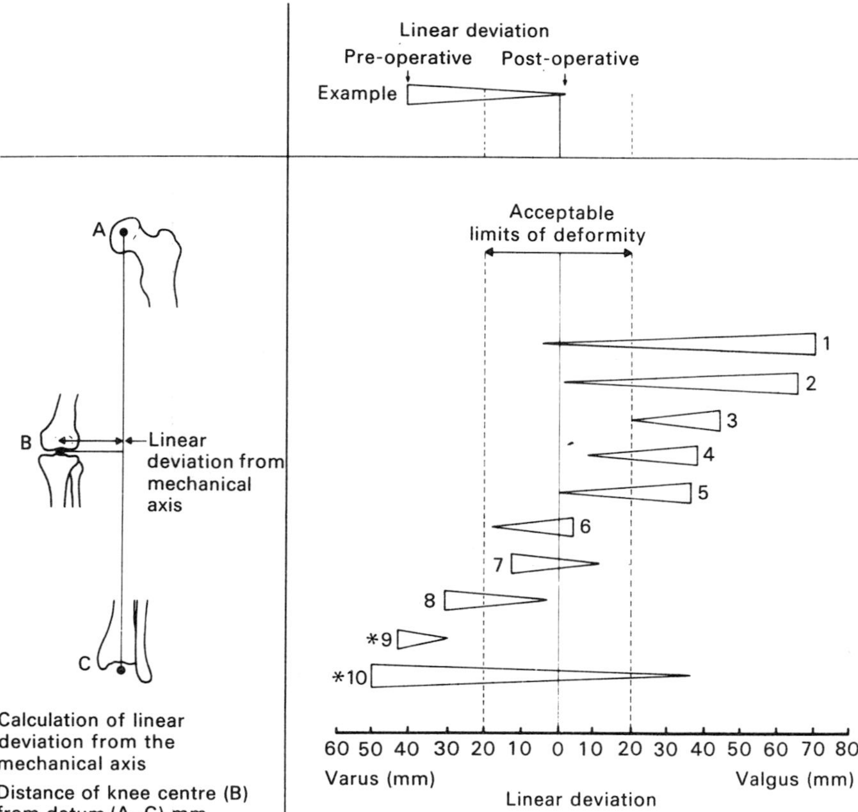

Fig. 32.2. Linear deviation of the knee: method of calculation and pre- and postoperative results from nine knees.

A sequence of 26 measurement points could then be taken from a pair of lower limbs in less than one minute. This helped to reduce errors through movement, and we have demonstrated this reliability in measurements taken at a six-month interval in our control knees in the knee arthroplasty study.

There is considerable advantage in the direct computer data acquisition, which saves time, reduces transcription errors, and eases subsequent data processing. This is of particular advantage in anthropometric studies.

The technique has advantages over radiographic techniques, in that it avoids the errors of recording data from two-dimensional films made from a point source of X-rays, particularly relevant to long leg films, and it does not involve irradiating the subject. Naturally, X-rays are needed in the assessment of the arthritic knee, but measurements of the whole leg deformity are obtained with the vector stereograph.

There are difficulties in the use of the vector stereograph in the patient with

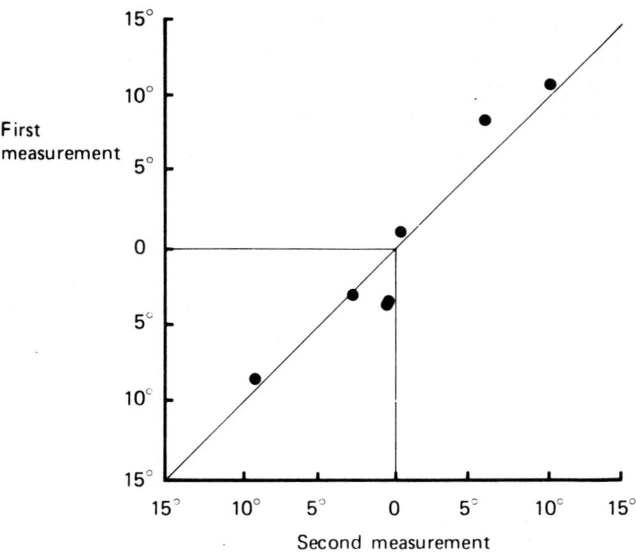

Fig. 32.3. A graph of repeat measurements of angular deviations of seven untreated knees separated by an interval of six months.

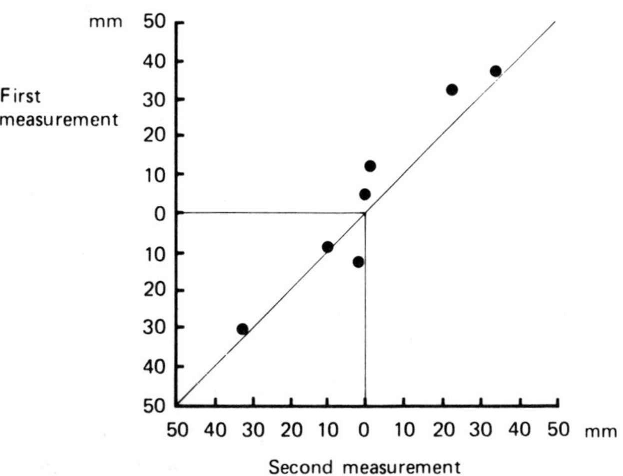

Fig. 32.4. A graph of repeat measurements of linear deviations of seven untreated knees separated by an interval of six months.

leg deformities. Many of the subjects were obese, and combined with joint destruction this made the identification of bony landmarks more difficult. However, many of the knees were cylindrical in section, making it easy to determine the midpoint of the joint. Our arthritic subjects had difficulty standing still during the measurements but the consistency of the repeat measurements of the control knees shows that this can be overcome. A further problem could arise from the nylon threads touching the subject, thus increasing their apparent length. In practice, with the measurement points we chose, this was not a difficulty.

It became apparent during the arthritis study that Freeman's datum point 3 cm medial to the anterior superior spine does not necessarily lie over the hip joint. It is more accurate in practice to take the mid-inguinal point as a landmark for the hip joint when inserting the prosthesis.

We present applications of the vector stereograph which have been of value to us in a clinical practice, and which we feel can be extended to further studies of this type. It may also be applied to measurement of the upper limb.

Acknowledgements

We wish to acknowledge the financial support of the East Anglian Regional Health Authority, the aid of the Medical Physics Department at Norwich, who constructed the vector stereograph, and the Oxford Orthopaedic Engineering Centre, who provided detailed engineering drawings.

REFERENCES

Clark, J. H. (1976). Designing surfaces in 3-D. *Comm. ACM* **19**, 454–60.

Fairbank, J. C. T., Pynsent, P. B., van Poortvliet, J. A., and Phillips, H. (1984). Mechanical factors in the incidence of adolescent patellofemoral pain. *J. Bone Jt Surg.* **66B**, 685–93.

Fuchs, H., Duran, J., and Johnson, B. (1977). A system for automatic acquisition of three-dimensional data. *AFIPS Nat. Comp. Conf.* **46**, 49–53.

Grew, N. D. and Harris, J. D. (1979). A method of measuring human body shape and movement—the 'vector stereograph'. *Engng Med.* **8**, 115–18.

Herron, R. E. (1972). Biostereometric measurement of body form. *Yearb. phys. Anthropol.* **16**, 80–121.

Kapandji, I. (1974). *The physiology of joints*, 2nd edn, Vol. 2. Churchill Livingstone, Edinburgh.

Leslie, I. J. and Bentley, G. (1978). Arthroscopy in the diagnosis of chondromalacia patellae. *Ann. rheum. Dis.* **37**, 540–7.

Lindström, K., Mauritzson, L., Benoni, G., Svedman, P., and Willner, S. (1982). Application of air-borne ultrasound to biomedical transducers. *Med. biol. Engng Comp.* **20**, 393–400.

Morris, J. R. W. and Harris, J. D. (1976). Three dimensional shapes. *Lancet* 1189–90.

Pynsent, P. B., Fairbank, J. C. T., Clack, F. J., and Phillips, H. (1983). Computer recording of anatomical points in three dimensional space. *J. biomed. Engng* **5**, 137–40.

Takasaki, H. (1975). Simultaneous all-round measurement of a living body by Moiré topography. *Photogrammetric Engng Remote Sensing* **41**, 1527–32.

33 Kinematic and Electrical Activity of Five Lower-Limb Muscles during a Fundamental Movement Pattern

C. L. HUBLEY

INTRODUCTION

It is essential that muscle function during human movement be described before muscle dysfunction can be completely understood. Identifying the eccentric, concentric, and isometric contraction sequences which occur during normal movement is an important component of rehabilitation and movement re-education programmes designed to produce normal muscle function. It has been demonstrated both on isolated muscle preparations (Bahler *et al.* 1968; Jewell 1982) and *in vivo* (Thorstensson *et al.* 1976; Komi 1973) that both the length of the muscle and the velocity of contraction affect the tension generated by the muscle and are thus important to the mechanics of the movement. A few *in vivo* studies have examined force production and absorption capabilities of selected muscles during different modes of locomotion using muscle kinematic data as their basis for description (Morrisson 1970; Grieve *et al.* 1978).

The purpose of this study was to use a non-invasive technique to describe muscle function of five lower-limb muscles during the performance of two types of vertical jump. The five muscles were the vastus lateralis, soleus, rectus femoris, gastrocnemius, and semimembranosus. The variables of interest were the muscle activation levels recorded using electromyography and the muscle kinematics using calculated muscle lengths and velocities from a simplified model of these muscles.

METHODOLOGY

Six male volunteers with no known pathological or neurological disorders were the subjects for this study. Subjects' height (cm), mass (kg), and age (years) were recorded prior to testing. Two vertical jumping conditions were examined, a counter movement jump (CMJ) in which the upward movement phase was preceded by a downward preparation phase and a static jump (SJ) which was performed from a static, squat starting position. Subjects performed two trials of each jump type while being filmed at 50 f.p.s. using a Locam camera placed perpendicular to the subjects.

Reflective low-inertia markers were placed over the anatomical landmarks necessary to calculate the lengths of vastus lateralis, rectus femoris, gastrocnemius, soleus, and semimembranosus (Fig. 33.1). The x–y coordinates of these anatomical landmarks were digitized, scaled, and corrected for parallax error and lens distortion on each frame of film. These coordinates were filtered by a Butterworth second-order, recursive, zero-lag, low-pass filter with a 4 Hz upper cut-off to remove high-frequency noise. These filtered absolute coordinates were used to calculate the joint angles necessary for modelling the muscle lengths. The muscle lengths were calculated by mathematical manipulation of the relative angular data, joint centres and the anthropometric data for that subject (Frigo and Pedotti 1978; Hubley 1981). Muscle velocities were calculated using finite differences of the displacement data.

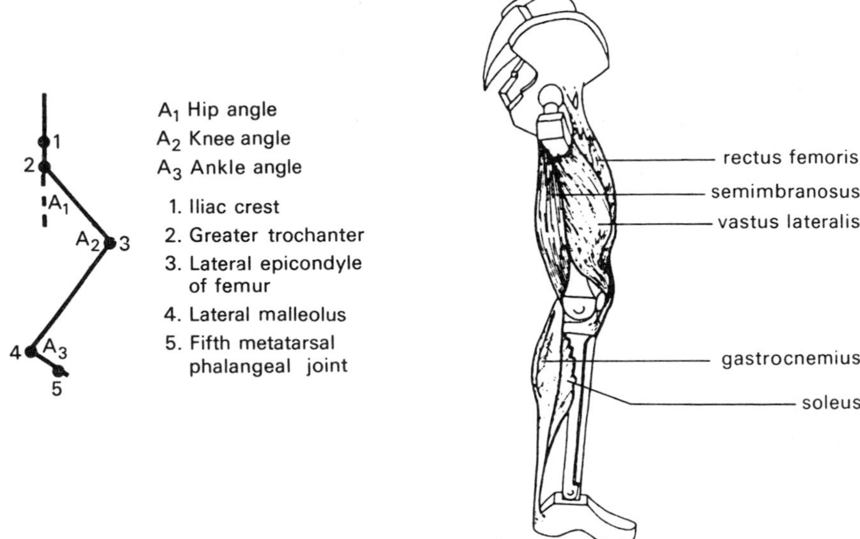

Fig. 33.1. Anatomical landmarks and joint angles used to calculate the lengths of the five lower limb muscles examined.

Beckman (16 mm) bipolar surface electrodes were placed 3 cm apart over the bellies of the five muscles of interest. A Conestoga Medical Electronics frequency modulated telemetry system (frequency response 30–300 Hz), mounted on a belt pack, transmitted the raw EMG signals from the subject to a Hewlett Packard analogue tape recorder (model No. 3968c) where they were stored for further processing. Prior to the test trials, subjects were requested to perform a maximal isometric contraction for each of the five muscles, by pushing against a resistance provided by an elgon table. The EMG signals were amplified, full-wave rectified and low-pass filtered at 6 Hz. These signals were A/D converted at 250 samples per second and the test trials were normalized to the maximum isometric values.

The EMG and film data were synchronized using a rectangular pulse generated by the camera. The EMG and muscle velocity data were plotted for each trial. Comparisons of these curves were made between the one and two joint muscles for the upward movement phase of the jump, which was defined as the period from when the velocity of the total body centre of mass became positive (UPM) to the point when the toes left the ground (TO).

RESULTS

General patterns across subjects were evident in four of the five muscles, the curves for the semimembranosus being variable between both subjects and jump types. Figures 33.2–33.5 give the EMG to muscle velocity relationships of four muscles for each of the two jump types. These curves are from one subject.

Figure 33.2 demonstrates the EMG velocity curves for vastus lateralis for the CMJ and SJ. Vastus lateralis differed between the two jump types as seen by the period of active lengthening prior to the UPM phase for the CMJ. However, from UPM to take-off the vastus lateralis actively shortened for both jump types.

Soleus data (the other single joint muscle) are presented in Fig. 33.3. The soleus muscle differed very little between jump types. It remained at the same length throughout most of the upward movement phase and shortened actively prior to take-off.

Both the rectus femoris and the gastrocnemius muscles (Figs. 33.4 and 33.5) demonstrated a period of active lengthening longer than that of either the soleus or vastus lateralis for both jump conditions. Only during the final stages of upward movement did those muscles demonstrate active shortening. The gastrocnemius shortened much later in the jump than the rectus femoris.

DISCUSSION

From these curves, the eccentric, concentric, and isometric contraction patterns of the four muscles can be determined for this simple movement pattern. It is evident from these curves that the major functions of the muscles change according to the phase of action within the movement. This information could not be obtained from either the joint angular kinematics or the EMG time-histories individually.

For example, the major function of the vastus lateralis was to contract concentrically during upward movement. The soleus muscle contracted isometrically during the majority of the upward movement phase and contracted concentrically prior to take-off, thus indicating that the soleus acted as a stabilizer prior to contributing to propulsion.

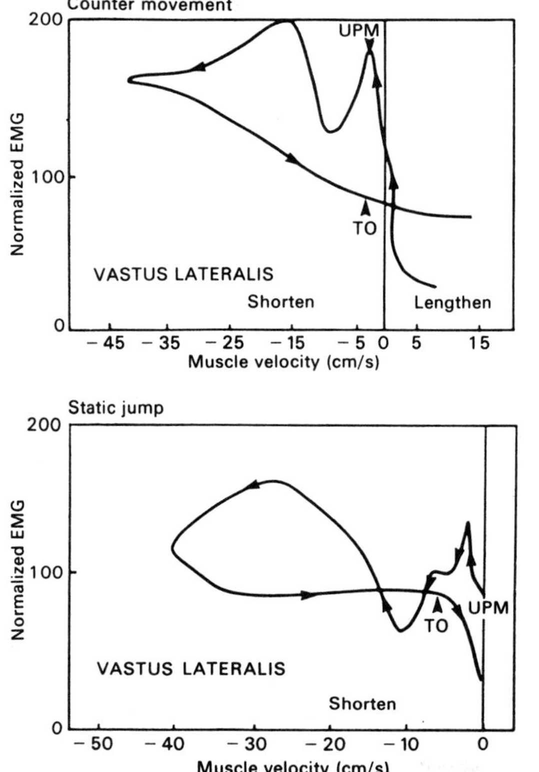

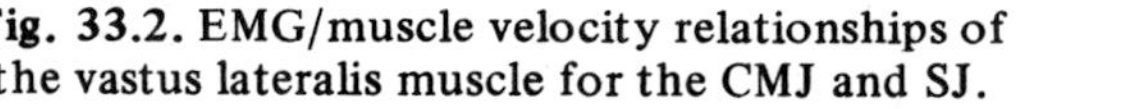

Fig. 33.2. EMG/muscle velocity relationships of the vastus lateralis muscle for the CMJ and SJ.

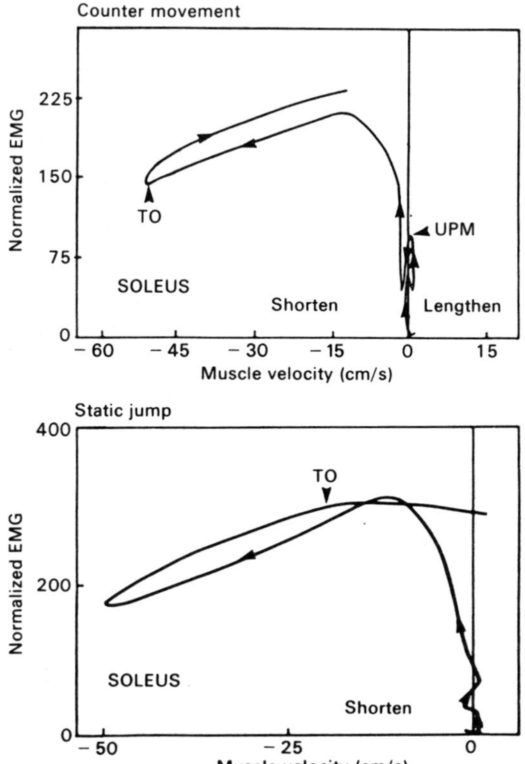

Fig. 33.3. EMG/muscle velocity relationships of the soleus muscle for the CMJ and SJ.

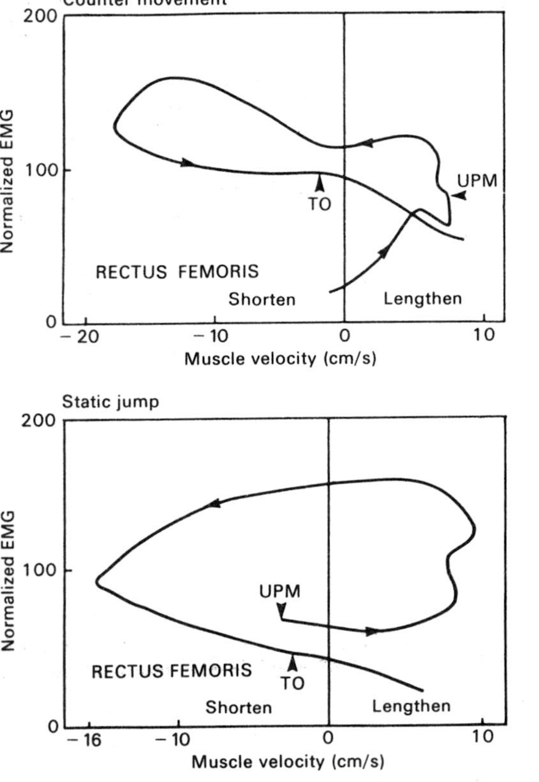

Fig. 33.4. EMG/muscle velocity relationships of the rectus femoris muscle for the CMJ and SJ.

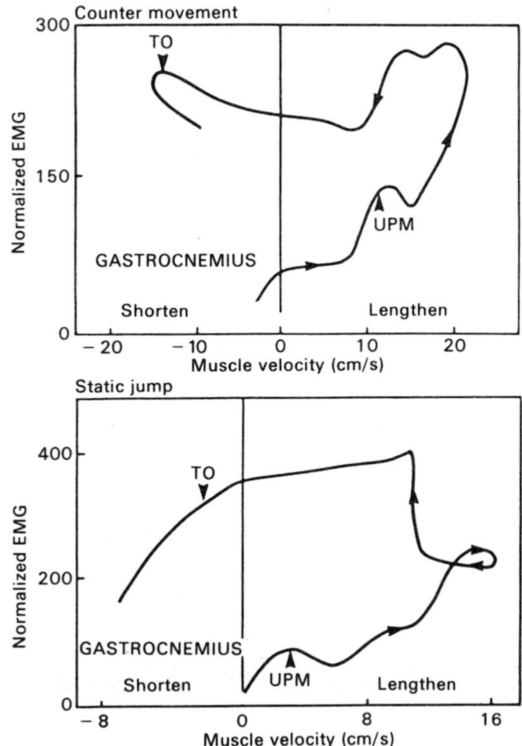

Fig. 33.5. EMG/muscle velocity relationships of gastrocnemius for the CMJ and SJ.

When comparing the function of the two joint muscles, the rectus femoris and gastrocnemius were eccentrically contracting for the majority of the upward-movement phase. However, both muscles did concentrically contract just prior to take-off to contribute to the total impulse. This pattern was similar between the two types of jump.

These results demonstrate that even for a basic movement which requires extensor patterns at the hip, knee, and ankle joints, the role of each extensor muscle may be quite different. Thus, the force production and absorption capabilities of these different muscles may not be accurately reflected by angular information, especially for the two joint muscles.

The technique used is relatively simple and could be easily modified for different kinematic data-collection systems. The information gained from such an analysis would be of value in the design of rehabilitation programmes which attempt to simulate normal muscle use. Also muscle kinematic data could be used in the calculation of individual muscle forces.

CONCLUSION

The kinematic and electrical activity of five lower-limb muscles have been presented, and used to describe muscle function during a fundamental movement pattern.

Acknowledgements

The author wishes to thank Dr R. P. Wells for his guidance and the Department of Kinesiology, University of Waterloo for use of their facility.

REFERENCES

Bahler, A. S., Fales, J. T., and Zierler, K. L. (1968). *J. gen. Physiol.* **51**, 369–84.

Frigo, C. and Pedotti, A. (1978). In *Biomechanics VI–A*, pp. 355–60. University Park Press, Baltimore.

Grieve, D. W., Pheasant, S., and Cavanagh, P. R. (1978). In *Biomechanics VI–A*, pp. 405–12. University Park Press, Baltimore.

Hubley, C. L. (1981). M.Sc. thesis, University of Waterloo, Ontario.

Jewell, B. R. (1982). In *Biomechanics VII–A*, pp. 75–83. University Park Press, Baltimore.

Komi, P. (1973). Medicine and sport. In *Biomechanics III*, pp. 224–9. Karger, Basel.

Morrison, J. (1970). *J. Biomech.* **3**, 51–61.

Thorstensson, A., Grimby, G., and Karlsson, J. (1976). *J. appl. Physiol.* **40**, 12–16.

34 Empirical Verification of a Linked-Segment Model of Rising from a Chair

L. L. LOWERY AND F. JOHNSON

INTRODUCTION

The study of joint replacement in people with arthritis includes a special interest in the loads encountered in various activities. Numerous biomechanical models have been developed for use in the analysis of joint forces and moments during various activities (Paul 1976; Morrison 1969; Ellis *et al.* 1979; Ramey and Yang 1981; Andriacchi *et al.* 1982). Morrison (1969) studied knee forces during the activities of level walking, walking up and down a ramp, and climbing and descending stairs. Andriacchi *et al.* (1982) studied knee flexion and extension moments for level walking and stair climbing. Ellis *et al.* (1979) examined knee forces whilst rising from a normal and motorized chair with and without the aid of arms.

The models mentioned so far are static analysis of motion. Predictions of dynamic moments have been studied by Thornton-Trump and Daher (1975), Chao and Rim (1973), Ramey and Yang (1981), Frank (1980), and Hemami and Jaswa (1978). These dynamic analyses were studied via computer simulation. The conventional methods of numerical filtering and differentiation techniques were used to obtain the kinematic information necessary for calculating the joint moments.

Except for a study of the dynamic control and stability of robots there are no other dynamic studies on rising from a chair reported in the literature. In walking, the variations of angles are small in contrast with the large angles produced when standing up. These large displacements of non-linearities of this motion are therefore very important, increasing the need for dynamic analysis of such movements. Although it is possible to estimate the joint forces from kinetic data (Morrison 1969; Ellis *et al.* 1979), it is not advisable for patients having painful joints to be used as subjects for any large number of trials. Thus, to find the optimal height of a chair seat for a particular disabled patient, one either uses experimental data derived from studies on normal people or finds another method.

This chapter suggests a combined simulation and empirical method which allows estimation of loads in the lower limbs. The method used is a linked-segment model to simulate the biomechanics of the process of rising from a chair.

THE MODEL AND THE METHOD

The model is a ten-segment symmetrical figure, with three moving segments; the tibia, femur and head, arms and trunk (upper seven segments moving as one). The anthropometric parameters are taken from Winter (1979). The computer program allows modification of the anthropometric data, segment length, weight, radius of gyration, and centre of gravity, for each subject. The three-linked model is given in Fig. 34.1. The moving segments are modelled in the

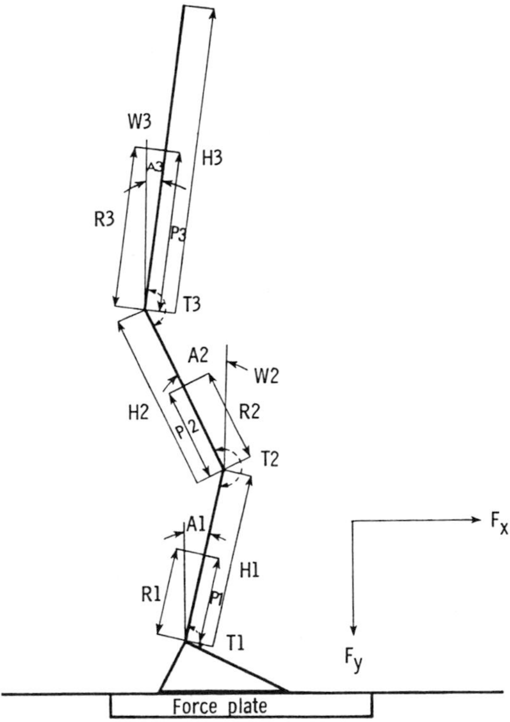

Fig. 34.1. The model showing the three links and the parameters measured and calculated.

sagittal plane. The angles of each segment to the vertical were measured by a three-channel polarized light goniometer (Crane Electronics). The outputs from the goniometer channels were digitized by the computer at a rate of 100 samples/second for two seconds of data or at a rate of 40 samples/second for five seconds of data collection. Before calculating velocity and acceleration the angles were filtered using a second-order low-pass Butterworth filter with a cut-off frequency of 5.5 Hz. The foot-ground reactions were measured with a six-component force plate (Kistler Ltd.). The forces due to chair contact were calculated by measuring the distances from the front edge of the chair to the knee joint (the distal end of the femur), the vertical chair height at the front edge of the chair, and the height of the hip joint from the floor. The weight on the chair was then calculated and the chair force estimated and added to the equations of motion, estimating the point of take-off from the chair and transfer of load to the floor.

The equations of motion for this model were derived by Hemami and Jaswa (1978) using Langrangian dynamics. These equations were adapted for a three-link model, adding the chair forces, to estimate the moments of the lower-limb joints. The moments were calculated from the upper end of the model, hip joint first followed by the knee joint and lastly the ankle joint. The moments are calculated independant of the ground-reaction forces although both are calculated from anthropometric and angular information. Appendix 34.1 gives the equations for the moments about the three segments shown in Fig. 34.1.

Data from nine female subjects were recorded to study the effects of various conditions to establish normal data. The subjects' ages were in the range 56–66 years. The subjects were asked to rise from the chair starting with a specific knee angle, chair height, arm position, and a speed according to their perception of fast, normal, and slow.

The ground-reaction forces and static and dynamic moments were computed and plotted against time, together with a reconstruction of the model position (Fig. 34.2). The model was then verified by comparing the calculated ground-

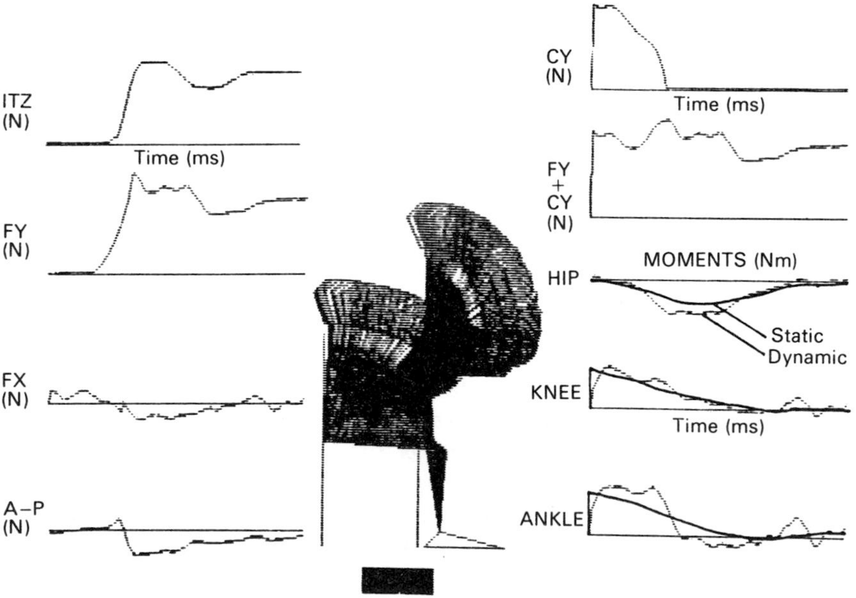

Fig. 34.2. The display of data and reconstruction of the model position.

reaction forces to those from the Kistler force-plate using the Spearman rank correlation method of statics. This gave correlation coefficients for the vertical and horizontal forces of 0.985 and 0.800, with significance levels of $p > 0.001$ and $p > 0.01$, respectively. Table 34.1 gives the correlation values of the reaction forces, peak forces of the vertical chair reaction forces, the measured and estimated vertical ground-reaction forces, and the trial velocity. The average velocities of the three segment angles for slow, normal, and fast were: 0.152 ± SD 0.021 radians/s; 0.294 ± SD 0.034; and 0.392 ± SD 0.023, respectively for all subjects.

Table 34.1. *Comparison of estimated and measured forces*

Spearman rank correlation coefficients of estimated v. measured forces				Vertical forces			Average velocity
Subject	Speed*	Vertical correlation	Horizontal correlation	Peak ITZ (N)	Peak Fy (N)	Peak Cy (N)	(radian/s)
1	F	0.975	0.643	612	620	693	0.332
	N	0.991	0.805	591	610	641	0.242
	S	0.999	0.823	563	561	536	0.122
2	F	0.979	0.789	696	727	753	0.415
	N	0.983	0.920	672	678	593	0.284
	S	0.994	0.831	627	640	623	0.186
3	F	0.983	0.909	775	745	710	0.401
	N	0.982	0.941	715	731	730	0.321
	S	0.974	0.950	789	773	630	0.183
4	F	0.923	0.700	932	967	888	0.363
	N	0.991	0.816	874	988	825	0.286
	S	0.975	0.685	825	828	771	0.153
5	F	0.948	0.765	787	809	822	0.348
	N	0.987	0.745	706	692	596	0.278
	S	0.980	0.678	684	655	639	0.100

*F = fast; N = normal; S = slow.

The calculation of area under the curve of joint moments shows that the knee moment is much higher for longer when the chair height is lowered, or the initial knee angle is increased to greater than 90°. The moment area is also greater for a slower rise (Table 34.2).

Altering all or some of these conditions can either increase or decrease the work needed to rise from a chair, depending on the individual's anthropometric data.

From these data on normal subjects the aim is now to produce a simulation of any individual, by entering anthropometric data specific to each subject.

Table 34.2. *Joint moment areas (Nms)*

Joint	Rate of rise					
	Fast		Normal		Slow	
	−	+	−	+	−	+
Ankle	−39.1	46.41	−30.32	78.12	−55.85	126.78
Knee	−8.14	44.42	−7.40	86.20	0.00	164.32
Hip	−82.50	0.00	−147.68	0.09	−208.16	3.03

The positive and negative joint moment areas are given for a subject weighing 56 kg.

It is hoped to predict the optimum conditions for getting out of a chair, for the individual subject.

CONCLUSIONS

The preliminary results given here demonstrate that a simple model which can be run on a small computer is capable of simulating the kinetics of a moving human body in the complex action of rising from a chair. By comparing direct recording of ground-foot reaction forces with those predicted by the model, an immediate verification of model performance can be obtained.

The implications of this work are that the measurement of the effects of changes in, for example, seat design may be determined by examination of the output from the computer. The model only needs to be adjusted to the anthropometry of the subject once, and a single verification performed. The model allows for estimates of the reaction between the seat and the person. Estimating the reaction forces between the arm of the chair and the upper limbs is an addition to be made which will permit the estimation of benefit to the patient in terms of reduction of upper-limb effort following possible lower-limb surgery, or improved seat design. This is an important consideration in the management of the severely disabled arthritic patient.

Acknowledgement

We are grateful to the Arthritis and Rheumatism Council for Research support of this work.

REFERENCES

Andriacchi, T. P., Galante, J. O., and Fermier, R. W. (1982). The influence of total knee replacement design and walking and stair climbing. *J. Bone Jt Surg.* **64A**, 1328–35.

Chao, E. Y. and Rim, K. (1973). The application of optimization principles in determining the applied moments in human leg joints during gait. *J. Biomech.* **6**, 499–510.

Ellis, M. I., Seedhom, B. B., Amis, A. A., Dowson, D., and Wright, V. (1979). Forces in the knee joint whilst rising from normal and motorized chairs. *Engng Med.* **8**, 33–40.
Frank, A. (1980). An approach to the dynamic analysis and synthesis of biped locomotion machines. *Med. biol. Engng* **8**, 465–76.
Hemami, H. (1978). Reduced order models for biped locomotion. *IEEE Trans. Syst. Man Cybernet.* **SMC-8**, 321–5.
— and Jaswa, V. (1978). On a three-link model of the dynamics of standing up and sitting down. *IEEE Trans. Syst. Man Cybernet.* **SMC-8**, 115–20.
Morrison, J. B. (1969). Function of the knee joint in various activities. *J. Biomed. Engng* **4**, 573–9.
Paul, J. J. (1976). Force actions transmitted by joints in the human body. *Proc. R. Soc. London* **192**, 163–72.
Ramey, M. R. and Yang, A. T. (1981). A simulation procedure for human motion studies. *J. Biomech.* **14**, 203–13.
Thornton-Trump, A. B. and Daher, R. (1975). The prediction of reaction forces from gait data. *J. Biomech.* **8**, 173–8.
Winter, D. A. (1979). *Biomechanics of human movements.* Wiley-Interscience, New York.

APPENDIX 34.1

The equations for the moments about the three segments shown in Fig. 34.1, given the following physical data:

n = Segment number.
An = Angle of segment n to the vertical.
Vn = Angular velocity of segment n.
Cn = Angular acceleration of segment n.
Wn = Mass of segment n.
Hn = Length of segment n.
Rn = Distance of centre of gravity from lower end.
Pn = Radius of gyration about centre of gravity.
Tn = Moment about lower end of segment.
Fx = Horizontal force at the floor.
Fy = Vertical force at the floor.
G = Acceleration due to gravity.
Tw = Total body weight.
K2 = 2 times W2 (two legs).
K1 = 2 times W1 (two legs).
PD = Distance from popliteal to edge of chair.
H = Length of floor to ankle joint centre.
Z = Weight of body segments in contact with the chair.
CY = Vertical force at the chair.
CX = Horizontal force at the chair.

Where a symbol is repeated, e.g. V3V3, the term is squared.

```
      IZ4=DBUF(J10+4)
      AP=DBUF(J10+5)
      LAT=DBUF(J10+6)
      ITZ=IZ1+IZ2+IZ3+IZ4
      IXPS=280.*(IZ2+IZ3)/ITZ
      DO 212 K=1,7
      A1L(K)=(AA1(J-4+K))/DIV
      A2L(K)=(AA2(J-4+K))/DIV
212   A3L(K)=(AA3(J-4+K))/DIV
      A1=A1L(4)
      A2=A2L(4)
      A3=A3L(4)
      V1=(A1L(7)-9.*A1L(6)+45.*A1L(5)-45.*A1L(3)
     1+9.*A1L(2)-A1L(1))/TI60
      C1=(2.*A1L(7)-27.*A1L(6)+270.*A1L(5)-490.*A1L(4)+
     1 270.*A1L(3)-27.*A1L(2)+2.*A1L(1))/TI2180
      V2=(A2L(7)-9.*A2L(6)+45.*A2L(5)-45.*A2L(3)
     1+9.*A2L(2)-A2L(1))/TI60
      C2=(2.*A2L(7)-27.*A2L(6)+270.*A2L(5)-490.*A2L(4)+
     1 270.*A2L(3)-27.*A2L(2)+2.*A2L(1))/TI2180
      V3=(A3L(7)-9.*A3L(6)+45.*A3L(5)-45.*A3L(3)
     1+9.*A3L(2)-A3L(1))/TI60
      C3=(2.*A3L(7)-27.*A3L(6)+270.*A3L(5)-490.*A3L(4)+
     1 270.*A3L(3)-27.*A3L(2)+2.*A3L(1))/TI2180
      V1V1=V1**2
      V2V2=V2**2
      V3V3=V3**2
      P=H1*COS(A1)+H2*COS(A2)+H
      FF2=-((F2/H2)*H2F*SIN(A2))
      ZMAX=(W3+FF2)*G
      Z=ZMAX*((PP2-P)/(PP2-PP1))
      IF(Z.LE.0.0)Z=0.0
      IF(Z.LE.0.0)ZMAX=0.0
      IF(Z.LE.0.0)FF2=0.0
      CY=Z-W3*(R3*V3V3*COS(A3)+R3*C3*SIN(A3))
      CX=-W3*(R3*C3*COS(A3)+R3*V3V3*SIN(A2))
      IF(Z.EQ.0.0)CY=0.0
      IF(CY.EQ.0.0)CX=0.0
      T(3)=C1*W3H1R3*COS(A1-A3)+C2*W3H2R3*COS(A2-A3)
     1         +C3*(W3P3P3+W3R3R3)
     1    -W3R3G*SIN(A3)
     1    +W3H1R3*V1V1*SIN(A3-A1)+W3H2R3*V2V2*SIN(A3-A2)
C
      T(2)=T(3)+C1*(W2H1R2+W3H1H2)*COS(A1-A2)
     1         +C2*(W2P2P2+W2R2R2+W3H2H2)+C3*W3H2R3*COS(A2-A3)
     1    -(W2R2G+W3H2G)*SIN(A2)
     1    -(W2H1R2+W3H1H2)*V1V1*SIN(A1-A2)+W3H2R3*V3V3*SIN(A2-A3)
D    1     +CY*H2*SIN(A2)
C
      T(1)=T(2)+C1*(W1P1P1+W1R1R1+W2H1H1+W3H1H1)
     1         +C2*(W2H1R2+W3H1H2)*COS(A1-A2)+C3*W3H1R3*COS(A1-A3)
     1    -(W1R1G+W2H1G+W3H1G)*SIN(A1)
     1    +(W2H1R2-W3H1H2)*V2V2*SIN(A1-A2)-V3V3*W3H1R3*SIN(A1-A3)
D    1     +CY*H1*SIN(A1)
      DO 10 K=1,3
      IF(T(K))1,2,2
1     ANEG(K)=ANEG(K)+T(K)*TI
      GOTO 10
2     APOS(K)=APOS(K)+T(K)*TI
10    CONTINUE
C
      FY=TWG-F1R1*(V1V1*COS(A1)+C1*SIN(A1))
```

```
      1 -F2H1*(V1V1*COS(A1)+C1*SIN(A1))
      1 -F2R2*(V2V2*COS(A2)+C2*SIN(A2))
      1 -W3H1*(V1V1*COS(A1)+C1*SIN(A1))
      1 -W3H2*(V2V2*COS(A2)+C2*SIN(A2))
      1 -W3R3*(V2V2*COS(A3)+C3*SIN(A3))
      1 -CY
      IF(FY.LE.1.0)FY=1.0
C
      FX=F1R1*(V1V1*SIN(A1)+C1*COS(A1))
      1 +F2H1*(V1V1*SIN(A1)+C1*COS(A1))
      1 +F2R2*(V2V2*SIN(A2)+C2*COS(A2))
      1 +W3H1*(V1V1*SIN(A1)+C1*COS(A1))
      1 +W3H2*(V2V2*SIN(A2)+C2*COS(A2))
      1 +W3R3*(V2V2*SIN(A3)+C3*COS(A3))
      1 -CX
      CALL MOVMAN(ANGLS)
      CALL MOMMAN(ANGLS,TMOMS)
      CALL COGMAN(ANGLS,ICOGX)
      ANGLS(1)=A1*57.3
      ANGLS(2)=-(A2-A1)*57.3
      ANGLS(3)=A3-A2*57.3
      V1=V1*R1
      V2=V2*R2
      V3=V3*R3
      V11=V1/R1
      V22=V2/R2
      V33=V3/R3
      SUM1=SUM1+V1
      SUM2=SUM2+V2
      SUM3=SUM3+V3
      SUM4=(SUM1+SUM2+SUM3)/3.
      SUM5=SUM5+V11
      SUM6=SUM6+V22
      SUM7=SUM7+V33
      SUM8=(SUM5+SUM6+SUM7)/3.
      N=IEND-IST
      P=(J-IST)/FLOAT(N)
      IF(P.LT..05)P=0.05
      IF(V1.GT.PK1)PK1=V1
      IF(V2.GT.PK2)PK2=V2
      IF(V3.GT.PK3)PK3=V3
      IF(ITZ.GT.PK4)PK4=ITZ
      IF(FY.GT.PK5)PK5=FY
      IF(T(1).GT.PK6)PK6=T(1)
      IF(T(2).GT.PK7)PK7=T(2)
      IF(T(3).GT.PK8)PK8=T(3)
      IF(T(1).LT.PK9)PK9=T(1)
      IF(T(2).LT.PK10)PK10=T(2)
      IF(T(3).LT.PK11)PK11=T(3)
      IF(CY.GT.PK12)PK12=CY
      DO 20 K=1,3
      IF(TMOMS(K))3,4,4
3     BNEG(K)=BNEG(K)+TMOMS(K)*TI
      GOTO 20
4     BPOS(K)=BPOS(K)+TMOMS(K)*TI
20    CONTINUE
      IF(CY.GT.PK14)PK14=TMOMS(1)
      IF(CY.GT.PK15)PK15=TMOMS(2)
      IF(CY.GT.PK16)PK16=TMOMS(3)
      IF(CY.LT.PK17)PK17=TMOMS(1)
      IF(CY.LT.PK18)PK18=TMOMS(2)
      IF(CY.LT.PK19)PK19=TMOMS(3)
```

35 Roentgen Stereophotogrammetric Analysis (RSA) in Total Hip and Knee Joint Replacement

G. SELVIK, L. I. HANSSON, A. LINDSTRAND, B. MJÖBERG, AND L. RYD

INTRODUCTION

Roentgen stereophotogrammetry is the term used for obtaining reliable three-dimensional measurements from radiographs. A specific system of roentgen stereophotogrammetric analysis (RSA) based on implantation of tantalum markers (tantalum balls with a diameter of 0.5 or 0.8 mm), highly accurate roentgen calibration equipment, and strict mathematical principles including rigid-body kinematic analysis has been clinically used at the University Hospital in Lund, Sweden, since March 1973 (Selvik 1974). By March 1983, 700 patients had been examined, mostly on repeated occasions, and close to 10 000 tantalum balls have been implanted without any reported complication. In orthopaedic practice the follow-up of bone-growth disorders of the lower extremity has dominated, with 161 patients investigated by March 1983. The main fields have been the treatment of leg length discrepancy (e.g. Bylander *et al.* 1983) and ankle fractures in children (e.g. Kärrholm *et al.* 1982). Also spinal fusions (47 patients, e.g. Olsson *et al.* 1977) and high tibial osteotomy for gonarthrosis (21 patients, Tjörnstrand *et al.* 1981) have been investigated. As a major field in orthopaedic surgery today is endoprosthetic replacement of the hip and knee joints, an interest has been focused to these areas and 86 and 98 patients, respectively, had been investigated by March 1983. As regards total hip protheses, methodological aspects and the follow-up of a limited number of patients were first presented by Baldursson *et al.* (1979, 1980), later followed by a study of surface replacement arthroplasty according to Wagner (Mogensen *et al.* 1982). At present all patients subject to primary total hip replacement at the Lund University Hospital are marked with tantalum balls at surgery and followed in a prospective study. The same has been done for almost two years for total knee joint replacement. This chapter will present findings from stress stereo-roentgen examinations (i.e. investigations with various external loadings) of hip and knee joint prostheses and compare these findings with information from standard radiography.

STRESS RSA IN TOTAL HIP REPLACEMENT

A study of 38 patients (41 hip replacements) who had pain and/or radiographic signs of loosening was performed. These patients were not from the prospective study and as they had not been marked with tantalum balls in their femur and pelvic bones at surgery; this was done 1–2 weeks prior to stereo examination using a percutaneous technique (Aronson *et al.* 1974). Tantalum balls cannot be introduced in the plastic acetabular component using this technique, and the centre of the circular indicator wire in the cup was thus determined, as well as the centre of spherical head of the femoral component. The cause for the replacement was mainly osteoarthrosis (36 hips), the prostheses used were Lubinus (20), Charnley (15), Brunswik (three), and Christiansen (three).

At stereo examination two X-ray tubes, 40° between the central rays, were used for simultaneous exposure. A reference plate with tantalum balls was placed in front of the two 30×40 cm films. Before patient investigations, a glass-plexiglass cage with tantalum balls at known positions was used as a calibration device. Without changing the positions of the X-ray tubes or the reference plate, the patient investigations were made. One pair of exposures was made with the hip loaded and one pair with the hip unloaded. Load was applied to the hip either by standing on the examined leg, or by the examiners compression of the hip in a longitudinal direction with the patient supine. The unloaded examination was performed either with the leg hanging freely or by distraction of the joint with the patient supine.

After measurement of the films in a photogrammetric instrument (Wild Autograph A8) supplied with television magnification, and data processing on a Univac 1100/80 computer, the displacements of the acetabular component in relation to the pelvis (represented by tantalum balls) and the head of the femoral component in relation to the trochanteric reference balls were found (Fig. 35.1).

Radiographical changes that may indicate loosening were considered to be present if there was:

1. A radiolucent zone between cement and bone of 2 mm or more.
2. A radiolucent zone between a prosthetic component and cement.
3. A fracture of cement or of the femoral component.

The accuracy of the method was evaluated by double examinations, performed because of insufficient information either on the acetabular or femoral component. Ten pelvic and 14 femoral double examinations were obtained. The standard deviations for the acetabular component were 0.29, 0.14, and 0.54 mm for the transverse (x), longitudinal (y), and sagittal (z) axis, respectively. The corresponding values for the femoral component were 0.12, 0.13, and 0.51 mm. Using Student's t-test the minimal significant translations ($p < 0.01$) were found to be 0.91, 0.44, and 1.72 mm for the acetabular component, and 0.35, 0.38, and 1.52 mm for the femoral component for the x-, y- and z-axis, respectively.

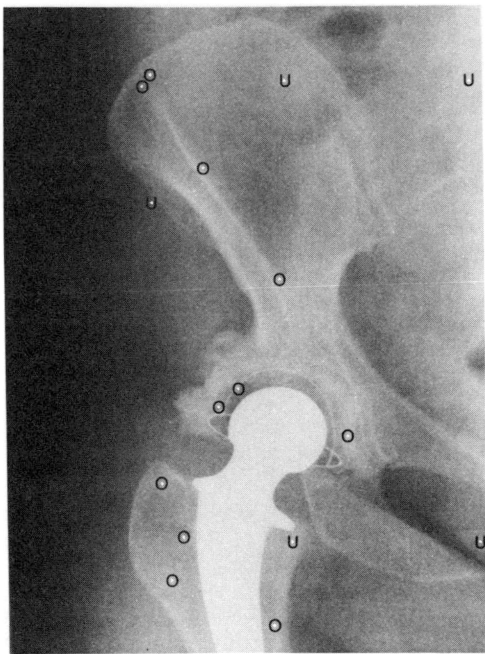

Fig. 35.1. One radiograph from a stereo pair showing tantalum markers in the pelvis (iliac bone), the femur (greater and lesser trochanter), and the acetabular cup. Reference markers in a plane in front of the film are indicated by a U. If no markers are present in the cup the centre of the circular wire is assessed.

Two acetabular components showed instability only along the x-axis, six only along the y-axis, and four along both the x- and y-axis; none showed significant instability along the z-axis. In three hips there was instability about 3 mm along the transverse axis, but there was no instability exceeding 1.2 mm along the longitudinal axis. Of the 12 hips with instability of the acetabular component only three showed corresponding radiographical changes, while seven acetabular components without instability showed radiographical changes (Fig. 35.2).

Four femoral components showed instability only along the x-axis, five only along the y-axis, seven along two or all three axes; none showed instability along the z-axis alone. Two femoral components were extraordinary unstable: in one case (IF_x = 9.3, IF_y = 6.1, and IF_z = 1.6 mm); a subsequent radiograph showed fracture of the femoral stem; in the other case (IF_x = 2.2, IF_y = 6.8, IF_z = 3.2 mm) the patient had Bechterew's disease. Out of the 16 hips with instability of the femoral component, 10 showed corresponding radiographical changes (RF), while 14 femoral components without instability showed radiographical changes (Fig. 35.2).

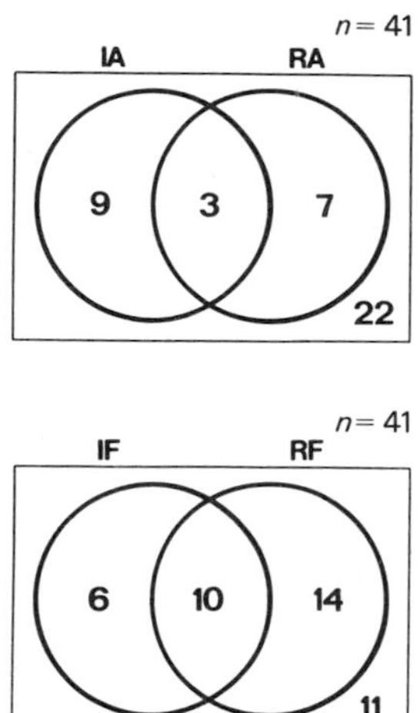

Fig. 35.2. Venn diagrams on coincidence of instability and radiographical changes for hip prostheses. IA and IF = instability of acetabular and femoral components, respectively. RA and RF = radiographical changes.

In summary, of 41 hips there was instability of the acetabular component in 12 and instability of the femoral component in 16, while there was no instability of any component in 14.

STRESS RSA IN TOTAL KNEE REPLACEMENT

Mechanical loosening of the tibial component is the single most frequent cause of complication in knee arthroplasty (Insall and Dethmers 1982). The influence of the radiolucent zone in conventional radiographs has been discussed in connection with mechanical loosening. The origin as well as the clinical importance of the zone is not fully understood. Micromovements have, however, been suggested as a cause as well as a consequence (Freeman *et al.* 1982) of this zone, which is known to consist of connective tissue, fibrous tissue or even fibrocartilage.

We have previously reported migratory movements of 20 out of 20 tibial components (Ryd *et al.* 1982). These migratory movements could, however, be caused by slow remodelling of the bone without the disintegration of the bone–cement interface. The objective of the present study is to investigate whether movements between the tibial component and the tibia in total knee arthroplasty could be induced by external loading. Furthermore we wanted to investigate if there would be a correlation between these possible movements and the size of the radiolucent zone.

Sixteen patients operated on by total knee replacement for osteoarthrosis had tantalum balls implanted at surgery in the tibia and the polyethylene tibial component. The postoperative follow-up period prior to this investigation was 1-3 years. At the time of the stress investigation all arthroplasties were rated as clinically successful and all the patients were satisfied.

A biplanar technique with the knee inside a calibration box was used for the stereo examination. The accuracy in this application has been found better than 0.1° for rotations and 0.06 mm for translations. In the stress investigation reference exposures were obtained with the patient in the supine position. Stress forces were produced by having a weight acting on the knee in different directions while standing on the investigated leg only. The following stress situations were analysed:

1. No external load applied.
2. Valgus, 10 kg.
3. Shearing, tibia pulled medially, 10 kg.
4. Shearing, tibia pulled forwards, 10 kg.
5. External rotation, patient standing on a rotation plate, with 5 kg applied to a 20 cm lever arm.
6. Internal rotation, as above.

The radiolucent zone was assessed in three different ways:

1. The maximum width of the radiolucent zone at any point in the interface was measured.

2. The fraction of the interface between the bone and the cement that had developed a zone was measured in the frontal and a lateral projections. The result, the zone extension, is given as the percentage of the entire interface.

3. The interface in the frontal and the lateral projections was subdivided into four parts each. The width of the zone in each of these eight subzones was measured and the values added together. This is called the zone value.

Since the objective of the study was to assess the magnitude of the maximum motion that can occur in ordinary interfaces, results are given as the maximum absolute movements found during the whole stress series. Only values exceeding 3 SD, that is 0.3° for rotations and 0.2 mm for translations, are considered significant.

There were significant rotations in 15 out of 16 prostheses. Thirteen prostheses rotated significantly about the transverse axis with a mean maximum of 0.5°. Thirteen prostheses rotated significantly about the longitudinal axis with a mean maximum of 0.6°; 10 prostheses significantly about the sagittal axis with a mean maximum of 0.4°. The mean of the maximum total rotation was 0.7°. No translatory movements were found for the centre of the prostheses.

Rotation of a rigid body, however, induces translatory movements out of the rotation axis. These movements in the periphery of the prostheses were analysed by calculating the translation (point motion) of each individual tantalum marker. The mean maximum for the 15 prostheses that showed instability was found to be 0.4 mm (range 0.2-0.7 mm). Assuming that the axis of rotation passes through the centre of the prostheses, the deflexions of the periphery were mathematically calculated to be 0.43 mm.

A radiolucent zone developed underneath all tibial components to a greater or lesser extent. Three prostheses showed a zone which exceeded 2 mm at its widest point, eight prostheses showed a zone between 1 and 2 mm and four had a zone of less than 1 mm. The mean zone extension was 49 per cent (range 20-100 per cent) and the mean zone value was 9 mm (range 2–18 mm). A statistically significant correlation was found between the magnitude of the instability and the size of the radiolucent zone both when the instability was expressed as rotations as well as point motion, and when the magnitude of the zone was expressed as a zone extension as well as the zone value (Fig. 35.3). The maximum width was, however, a less relevant way to assess the radiolucent zone.

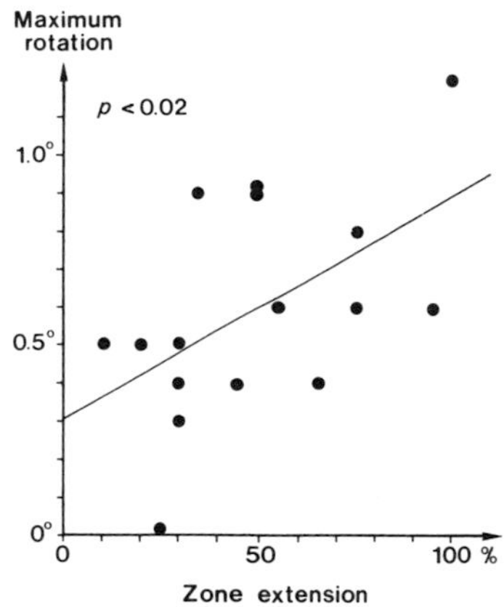

Fig. 35.3. Linear regression analysis of maximum rotation versus zone extension at stress RSA of knee prostheses. Coefficient of correlation = 0.538, n = 16.

In three patients, two consecutive stress series were performed to assess the reproducability of the induced movements, and the overall SD for the different types of stress was found to be 0.15°.

DISCUSSION AND CONCLUSIONS

According to the *Concise Oxford Dictionary* the term 'loose' denotes 'detached or detachable from its place'. Both aspects of loosening can be studied using RSA. The former aspect can conveniently be termed migration (displacement with time), the latter instability (displacement with load). RSA was found to be practical and accurate in revealing instability of hip and knee joint endoprostheses. The correlation between RSA-verified instability and signs of loosening by conventional radiography was demonstrated and found to be bad for the hip but good for the knee.

For total hip prostheses, conventional radiography is particularly uncertain in regards to the acetabular component, with reported rates of loosening varying from 1 to 29 per cent in five-year follow-up studies. This was clearly demonstrated also in the present study where 19 hips showed either RSA or radiographic evidence of loosening, but only three fulfilled both criteria. Reasons for underestimation using RSA can either be too high significance limits for displacements (the present limits were based on re-examinations of cases with unfavourable photogrammetric criteria) or too small loading. Either a lower limit or a higher stress might have demonstrated movement. However, it is well demonstrated that conventional radiography is uncertain, as nine hips with RSA-verified instability were not detected.

Numerous authors have reported the occurrence of a radiolucent zone underneath the tibial component in knee arthroplasty. Neither the cause nor the consequences of this radiolucent zone is fully understood and for practical clinical purposes a zone of less than 2 mm in width has come to be regarded as 'normal' and acceptable. The present study indicates that movements occur between the bone and the prosthesis in most knee arthroplasties in response to physiological forces. There were no radiographic indications of loosening between the prosthesis and the cement, in terms of cement fractures or radiolucencies within the cement itself. This fact and clinical experience makes it probable that the instability in this study occurs between the bone and the cement.

The mean instability found was 0.7°, which corresponds to a maximum point motion along the periphery of 0.4 mm. This latter value represents the maximum deformation that takes place within the soft tissue of the radiolucent zone. Considering the fact that the radiolucent zone is 1-2 mm in width this instability appears to be within a reasonable order of magnitude. The reproducability of the movements was within 0.15°. Included in this figure is the error of the method itself (0.1°). This high reproducability suggests a rather high degree of mechanical organization within the zone tissue. The behaviour of the prostheses in this study could be explained by elastic properties of the radiolucent zone.

The results of knee arthroplasty are becoming better and better, with frequencies of excellent or good exceeding 90 per cent (Insall *et al.* 1979) in short term follow-up. The long-term results are however worse, and not very many arthroplasties survive 10 years (Tew and Waugh 1982). Among causes of failure are infection, mediolateral instability, patellar malfunctioning, and prosthetic loosening. The patient with a failing knee arthroplasty therefore presents with a variety of symptoms, the cause of which could be one of many alternatives.

This study has shown that instability is not synonymous with loosening as regards the tibial component, but can be explained by elasticity of the fibrous zone. The higher resolution of RSA at the knee as compared to the hip, and the possibility to determine rotations, since markers were implanted in the polyethylene, are reasons for this added information. The situation might be the same for the acetabular component, and by the current marking of the cup it should be possible to answer this question in the future.

The specific value of RSA will probably be its ability to detect instability and migration at an early stage and thus evaluate new techniques for total joint replacement. Also, RSA is valuable in identifying loosening among cases of clinical failure. This is of importance, since radiographical changes appear late in the course of loosening and early signs indicating this complication have not been described. However, as methods now have been refined it is necessary to assess the limit between normal and pathologic instability.

REFERENCES

Aronson, A. S., Holst, L., and Selvik, G. (1974). An instrument for insertion of radiopaque bone markers. *Radiology* **113**, 733–4.

Baldursson, H., Egund, N., Hansson, L. I., and Selvik, G. (1979). Instability and wear of total hip prostheses determined with roentgen stereophotogrammetry. *Archs Orthopaed. Traumat. Surg.* **95**, 257–63.

—, Hansson, L. I., Olsson, T. H., and Selvik, G. (1980). Migration of the acetabular socket after total hip replacement determined by roentgen stereophotogrammetry. *Acta orthopaed. scand.* **51**, 535–40.

Bylander, B., Hansson, L. I., and Selvik, G. (1983). Pattern of growth retardation after Blount stapling: a roentgen stereophotogrammetric analysis. *J. pediat. Orthopaed.* **3**, 63–72.

Freeman, M. A. R., Bradley, G. W., and Revell, P. A. (1982). Observations upon the interface between bone and polymethylmethacrylate cement. *J. Bone Jt. Surg.* **64B**, 489–93.

Insall, J. N. and Dethmers, D. A. (1982). Revision of total knee arthroplasty. *Clin. Orthopaed.* **170**, 123–30.

—, Scott, W. N., and Ranawat, C. S. (1979). The total condylar knee prothesis: a report of 220 cases. *J. Bone Jt Surg.* **61A**, 173–80.

Kärrholm, J., Hansson, L. I., and Selvik, G. (1982). Roentgen stereophotogrammetric analysis of growth pattern after supination—eversion ankle injuries in children. *J. pediat. Orthopaed.* **2**, 25–37.

Mogensen, B., Ekelund, L., Hansson, L. I., Lidgren, L., and Selvik, G. (1982). Surface replacement of the hip in chronic arthritis. A clinical, radiographic and roentgen stereophotogrammetric evaluation. *Acta orthopaed. scand.* **53**, 929–36.

Olsson, T. H., Selvik, G., and Willner, S. (1977). Mobility in the lumbosacral spine after fusion studied with the aid of roentgen stereophotogrammetry. *Clin. Orthopaed.* **129**, 181–90.

Reckling, F. W., Asher, M. A., and Dillon, W. L. (1977). A longitudinal study of the radiolucent line at the bone-cement interface following total joint replacement procedures. *J. Bone Jt Surg.* **59A**, 355–8.

Ryd, L., Lindstrand, A., and Selvik, G. (1982). Micromovement of the tibial component in successful knee arthroplasty, studied by roentgen stereophotogrammetry. In *Biomechanics: principles and applications* (ed. R. Huiskes, D. H. van Campen and J. R. de Wijn). Martinus Nijhoff, The Hague.

Selvik, G. (1974). A roentgen stereophotogrammetric method for the study of the kinematics of the skeletal system. Thesis, Lund.

Tew, M. and Waugh, W. (1982). Estimating the survival time of knee replacements. *J. Bone Jt Surg.* **64B**, 579-82.

Tjörnstrand, B., Selvik, G., Egund, N., and Lindstrand, A. (1981). Roentgen stereophotogrammetry in high tibial osteotomy for gonarthrosis. *Archs Orthopaed. Traumat. Surg.* **99**, 73-81.

36 Applications of a Microcomputer-Controlled Hand-Assessment System

A. R. JONES, A. UNSWORTH,
AND I. HASLOCK

INTRODUCTION

Hand assessment is necessary so that clinicians and therapists can evaluate how a patient is responding to a specific treatment or how a disease is progressing. This can cover any treatment from surgical intervention to manipulation to improve the functional ability in an arthritic patient. With the aid of assessment, the effectiveness of treatment can be judged and the point determined where the patient is fit enough to return to normal activity.

To be successful, a method of assessment needs to be quick, accurate, reproducable, and simple to use. With these attributes there is no time wasted, the patient is not subjected to fatigue, and the results will be meaningful.

Many hand assessments have been published of which Carthum *et al.* (1969), Jebsen *et al.* (1969), MacBain (1970), and Smith (1973) are typical examples. From these, and other published papers, it appears that most centres use their own system and there is little standardization.

If no quantitative assessment routine is used, the patient is assessed on typical activities of daily living (ADL). This is mainly subjective and is based on the experience of the assessor.

Some assessment schemes try to achieve some objectivity by timing the tasks being carried out, and obtaining a functional index, that is supposed to relate to the state of the hand.

Another disadvantage of many schemes is complexity, owing to the many tasks that are included. This leads to a long period of testing resulting in patient fatigue.

From this background it was thought necessary to develop a scientifically based automatic hand function assessment system, which was simple to use by non-technical staff, was robust, reliable, and would provide results that could be immediately interpreted by the operator.

APPARATUS

Measuring devices

The devices used measured function and strength by means of strain-gauged transducers.

Isometric strength of individual fingers and the total grip exerted

This consisted of a heel plate, shaped to fit comfortably into the heel of the hand, and a key plate, containing four transducers (Fig. 36.1). These measured the individual finger strength when a maximal grip was exerted, between the key and heel plates. Total grip strength was the addition of the four finger strengths at every point in time.

Pan-handle grip and lifting force

A transducer in a pan-handle measured the grip force as the pan was lifted (Fig. 36.1). The movement of the pan was limited by a twin-rate spring attached to another transducer which measured the lifting force.

Kettle-handle grip and lifting force

The kettle-handle, complete with a grip-measurement transducer, was fitted on top of the pan thereby utilizing the same lifting-force transducer. Again the grip was measured as the handle was lifted.

Isometric key twist

The amount of torque that could be applied in either pronation or supination was determined by attempting to turn a simulated key. This was fixed to a dual-sensitivity torquemeter. The key could also be fixed at various angles of pronation or supination (in increments of 15°).

Isometric pinch strength

This was measured by squeezing together two small platens attached by a pivot arrangement to a force transducer. Both lateral and individual finger pulp pinch were measured. Two transducers of different sensitivities were provided.

Both the key and pinch devices were located in a single unit (Fig. 36.1) which was designed by Robertson (1981) to investigate the variation of torque and pinch force with pronation and supination.

Isometric extensor lift strength

This used the fine-sensitivity pinch-force transducer. A flat board was placed underneath the lower platen, with the hand placed palm down on the board. With the finger nail of the required finger situated underneath and just touching the lower platen, the subject was required to extend the finger against the platen. The other fingers were gently restrained from lifting at the same time.

Isometric tube twist

A 30 mm diameter, 150 mm long tube replaced the key on the torquemeter. The subject was then required to attempt to turn it away from him as if it were a cloth being wrung out.

Fig. 36.1. View of the complete hand-assessment system showing the computer (A), pan- and kettle-handle (B), grip (C), and key and pinch transducer (D).

Electronic circuitry

The device transducers were multiplexed into a single strain-gauge amplifier via a conditioning unit. The amplifier output signal was sent to an eight-bit analogue-to-digital converter (ADC) that converted the analogue voltage to its digital equivalent number between 0 and 255.

Communication between the circuitry and computer was by a digital interface (DI–09, supplied by Data Efficiency, Hemel Hempstead) that plugged directly into the rear of the microcomputer.

Microcomputer

The microcomputer used in this work was an Apple II Europlus microcomputer, programmed to control all the circuitry, and to collect, display and store all the device data, as required by the operator.

For each patient the operator was required to type in the patient's name, test number, date of birth, and dominant hand. The computer then displayed a device menu from which the operator selected a device. The computer switched in the circuitry for the selected device and prompted the operator with the data-collection instructions. Figure 36.2 gives a printout showing the interactive instruction nature of the program. During collection the computer was sending

```
PATIENTS' TEST NUMBER CONSISTS OF
            #######-*

WHERE ####### = HOSPITAL RECORD NUMBER
  AND    * = PATIENTS' ASSESSMENT NUMBER

TEST NUMBER?-->
PATIENTS' DATE OF BIRTH?
-AS DY/MT/YR-->
PATIENTS' DOMINANT HAND?L/R-->
DEVICE CODES ARE:-

          GRIP----------------GR
          PAN & KETTLE HANDLE-PH & KH
          F & C LATERAL PINCH-FLP & CLP
          F & C PULP PINCH----FPP & CPP
          F & C KEY TWIST-----FKT & CKT
          L & S TUBE TWIST----LTT & STT
          FINGER LIFT---------EX

OR CONTROL CODES:-

          STOP----------------S
          RECALIBRATE---------R
          NEW PATIENT---------P

TYPE IN CODE FOR DEVICE REQUIRED?GR
GRIP DEVICE
MAX DATA COLLECTION TIME IS 10 SECONDS



          EVERYTHING IS READY

DEPRESS EITHER RED START BUTTON TO
           INITIATE SCAN
KEEP DEPRESSED AND RELEASE WHEN COMPLETE

SCAN COMPLETED

CALCULATIONS PROCEEDING

I'M SORRY THIS WILL TAKE ME A SHORT TIME
INDEX FINGER MAXIMUM= 112.6 NEWTONS @ 1.22 SECONDS
MIDDLE FINGER MAXIMUM= 145.32 NEWTONS @ 1.31 SECONDS
RING FINGER MAXIMUM= 130.62 NEWTONS @ 1.41 SECONDS
LITTLE FINGER MAXIMUM= 56.87 NEWTONS @ 1.5 SECONDS
TOTAL MAXIMUM GRIP= 445.41 NEWTONS @1.5 SECONDS

IS DATA TO BE STORED ON DISC?Y/N-->
```

Fig. 36.2. Computer printout of interaction obtained during a measurement.

out and receiving control pulses for the ADC and switching transducer channels as necessary. For example, in a grip-strength measurement, each transducer was 'ready' sequentially over the grip measurement.

When collection was complete, the computer performed the necessary calculations, the results of which it then displayed in graphical form on the monitor. Maximum values for each channel were also displayed and the operator decided whether or not to store the results on floppy disc. Figure 36.3 shows a summary flowchart of the data-handling routine.

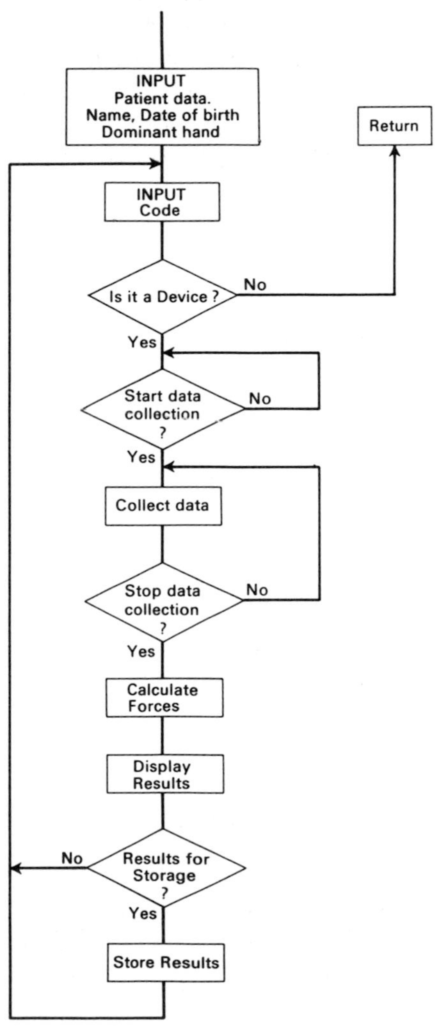

Fig. 36.3. Flowchart of data-handling routine.

Calibration

All the force-measuring devices were calibrated, using an Instron 3522 tensometer to apply loads directly to the operating surfaces of each device. The twist devices had a lever attached to the transducer from which weights were hung.

The calibration was checked using the Instron to apply a range of known loads several times each. Table 36.1 shows the mean percentage discrepancy of the devices.

Table 36.1. *A summary of device reproducibility*

	Device	Reproducibility (%)
Grip	Index	5.0
	Middle	3.1
	Ring	5.3
	Little	−1.4
Pan	Grip	−0.7
Kettle	Grip	−0.7
	Lift 1	1.5
	Lift 2	−3.3
Fine key-twist		5.3
Coarse key-twist		4.2
Fine pinch		−0.9
Coarse pinch		2.3

Repeatability

Four normal subjects were asked to use the devices several times with alternate hands in a continuous testing session. Table 36.2 gives the coefficient of variation obtained for each device. Individual finger strength had a large coefficient of variation of between 6.0 per cent and 33.0 per cent for different subjects; total finger strength had a range of 5.4 per cent to 12.1 per cent with a mean of 7.8 per cent.

The pan-handle grip and lift variation were also quite high at 14.4 per cent and 12.8 per cent respectively.

The coefficients of variation for the key-twist and pinch devices had previously been found to be 4.3 per cent.

METHODS

The patients were volunteers from various outpatient clinics within the hospital. All of these were tested for grip-strength, pan- and kettle-grip and lift, pulp and

Table 36.2. *Coefficient of variation of the measuring devices*

	Device	Mean coefficient of variation (%)
Grip	Index	15.4
	Middle	17.5
	Ring	14.7
	Little	12.9
	Total	7.8
Pan	Grip	14.4
	Lift	12.8
Key-twist and pinch		4.3

Coefficient of variation = $(\sigma/\bar{x}) \times 100$ per cent.

lateral pinch, key- and tube-twist, and extensor strength. Owing to limitations of time only one measurement per device per hand was carried out.

Patients were studied from three sources. Patients with early rheumatoid arthritis who had just started attending a penicillamine-auranofin drug trial were monitored at monthly intervals. Patients attending the physiotherapy clinic for a wide range of hand ailments were monitored at fortnightly intervals. Both sets of patients were tested to see if the prescribed drug or physiotherapy treatment had any effect on their strength results. Long-term arthritic patients, attending the rheumatology clinic for gold injections were monitored at monthly intervals.

For all tests, the patients were invited to try each device with each hand prior to measurement. Spoken encouragement, i.e. to squeeze, grip, lift or twist as hard as they could, was only given prior to a measurement, with nothing being said actually during the test.

Grip strength, lateral pinch, key- and tube-twist were measured with the patient seated. The devices were set at a height such that the patient's arm was horizontal. For measurements of grip strength, the elbow joint was held at 90° of flexion by resting the forearm on the arm of a chair with the device held vertically.

The patients were required to stand to use the pan, kettle, and pulp-pinch devices. The pan and kettle were fixed at the height of a typical cooker or kitchen work surface.

In all tests the patient's hand was allowed to be in 'ready' contact with the device prior to measurement. Measurement time was variable but usually in the range of 4 to 6 s. Pinch and extensor measurements took 2 s per finger.

RESULTS

Patients have been followed at regular intervals during their treatment. To illustrate the use of the equipment the results of one patient undergoing a penicillamine-auranofin drug trial are presented below.

Isometric grip strength (Fig. 36.4)

The total grip had an initial drop from 89 N to 69 N, but over subsequent weeks gradually improved to 103 N. The individual finger strength showed a lot of variation but a similar trend to the total strength can be seen.

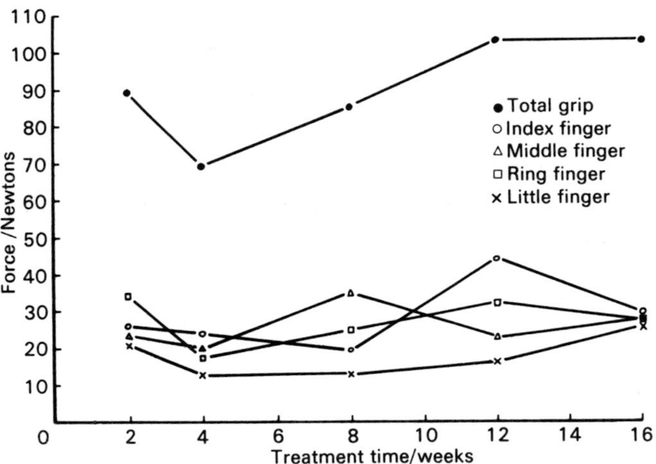

Fig. 36.4. The variation of individual finger and total grip strength with time of a patient undergoing drug treatment.

Pan-handle and kettle-handle (Fig. 36.5)

Both devices followed a similar pattern. Both lifting forces started at around 5 N but subsequently the kettle lift showed the largest increase rising to 31 N before falling to 22 N at week 16, the pan lift subsequently rose to 18 N and then fell to 14 N in the same period.

The pan grip was stronger than the kettle grip, starting at 26 N, falling at week 4 to 19 N before a gradual rise to around 36 N. The kettle was gripped initially with a force of 15 N, falling over the next two visits to 2 N but then rising finally to 12 N.

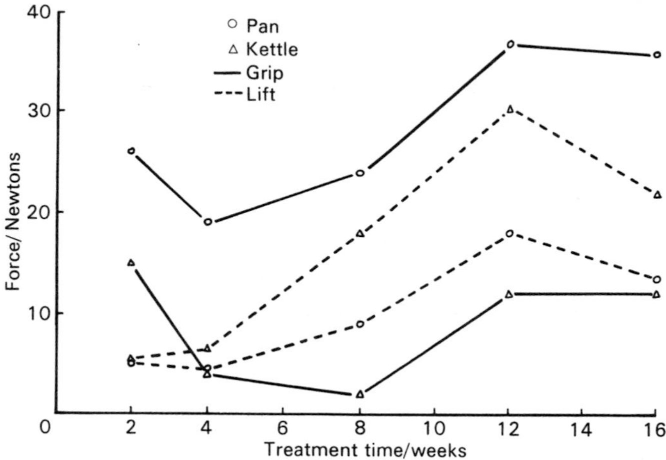

Fig. 36.5. The variation of pan- and kettle-handle grip and lift with treatment.

Key and tube-twist (Fig. 36.6)

The key-twist showed a gradual increase from 0.44 Nm to 0.92 Nm over the 16 weeks of treatment.

The tube-twist also showed an improvement, but this was a step change from around 1.6 Nm to around 3.2 Nm between the eighth and twelfth week of treatment.

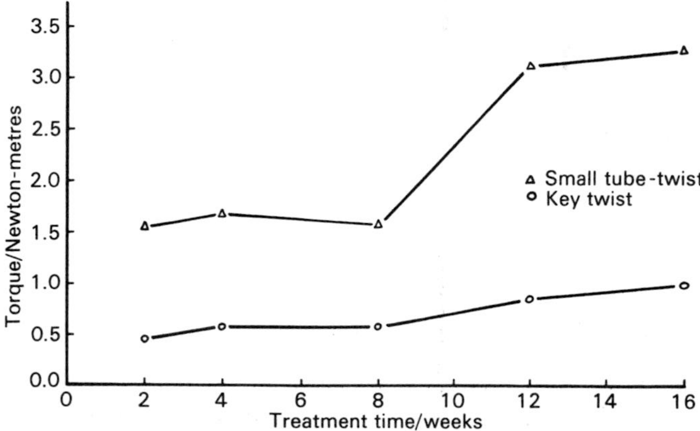

Fig. 36.6. The variation of key and tube-twist torque with treatment time.

Pulp-pinch (Fig. 36.7)

The index and middle fingers had the largest pinch forces, both rising gradually from around 14 N to around 23 N on the final visit.

The ring and little fingers had the same pinch initially (6 N) but these separated with subsequent measurements. They both increased to around 11 N on the eighth week and dropped on the next visit to around 7 N. On the final visit, however, the little finger rose to 11 N while the ring finger increased to 22 N.

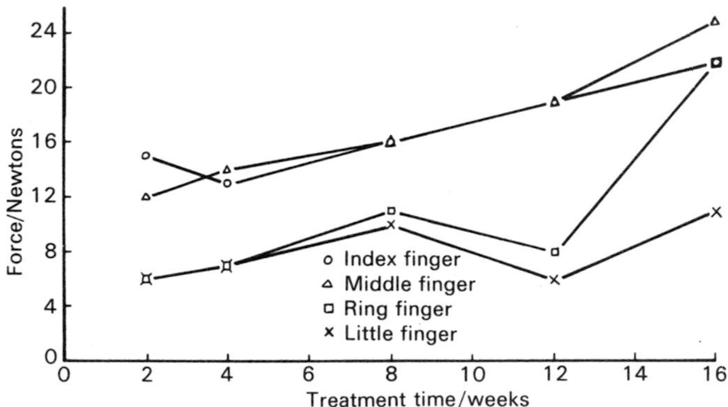

Fig. 36.7. The variation of individual finger pulp-pinch strength with treatment.

Extensor lift (Fig. 36.8)

All the fingers were in the range of 1.3 N to 4.1 N and showed a great deal of variation, with no obvious pattern of finger lift strength.

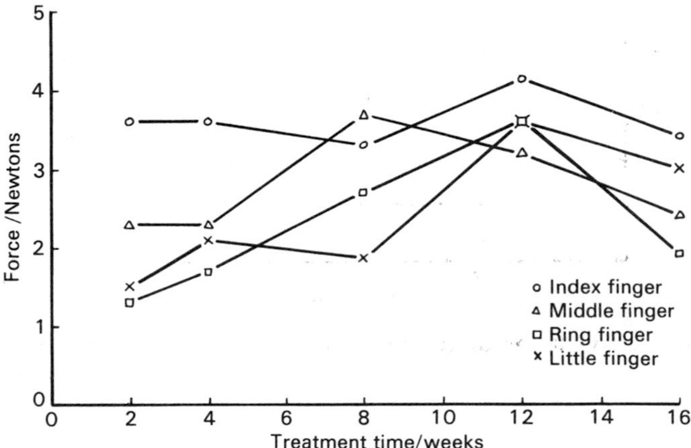

Fig. 36.8. The variation of individual finger extensor force with treatment.

DISCUSSION

The computer-controlled hand-assessment system described here produces a large number of results both quickly and accurately. The operation can be fairly time-consuming when testing patients on all the devices, but patient fatigue appears not to be a problem. This is due to a relatively long resting time between tasks each of which only takes a few seconds to perform.

Reproducability of the devices is shown to be good, all within ±5.3 per cent, but some repeatability tests on individual fingers of individual subjects appear to be higher, 33 per cent maximum for the coefficient of variation of the middle finger strength. The high finger strength variation possibly reflects the large number of ways of gripping. However, even with a high individual finger strength variation the total grip variation is acceptable at 7.8 per cent.

The results also show that the test results on individual finger measurements appear to have a larger variation than the test results of whole hand measurements. This appears to confirm the high coefficient of variation results obtained in the finger strength repeatability tests (Table 36.2) and indicates that a better measurement of the patient's progress is possibly obtained from full hand testing.

The computer software has been produced to provide a fully instructive interaction with the operator. This ensured that the procedure was easy to follow and suitable for use by non-technical staff. The facilities of the computer enabled a full graphical display of the results, making for quick recognition and understanding of the trends. Disc storage of these results meant that minimum information needed to be printed out during the tests. A full numerical printout could be obtained after the test for a fuller examination of the results.

Results from the system have generally indicated a performance change over the time of treatment. Full analysis of the results is still proceeding and will be published subsequently.

CONCLUSIONS

The computer-controlled hand-assessment system has been shown to be sufficiently accurate and simple to use, and it produces negligible patient fatigue. It obtains results quickly and in a graphical form which can easily be understood. A numerical printout for the patient's notes can be obtained after the tests.

In the drug trial results demonstrated here, the patient was seen to be improving in hand function throughout the period of examination. This illustrates the usefulness of such a quantitative method of assessment in monitoring a patient's response to treatment.

Acknowledgement

The authors would like to thank Mr Ray Mand for manufacturing the measuring devices, Mrs Valerie Rhind, and the physiotherapy and rheumatology staff of

Middlesbrough General Hospital for their assistance in providing patients for the study, and Mrs Christine Wright for typing the manuscript.

This work was financed by the DHSS and the authors would like to acknowledge their support.

REFERENCES

Carthum, C. J., Clawson, D. K., and Decker, J. L. (1969). Functional assessment of the rheumatoid hand. *Am. J. occup. Ther.* **32**, 122–5.

Jebsen, R. H., Taylor, N., Trieschmann, R. B., Trotter, M. J., and Howard, L. A. (1969). An objective and standardised test of hand function. *Archs phys. Med. Rehab.* **50**, 311–19.

MacBain, K. P. (1970). Assessment of function in rheumatoid arthritis. *Can. J. Occup. Ther.* **37**, 95–102.

Robertson, A. (1981). A study of human key grip. A final years honours project, University of Durham.

Smith, H. B. (1973). Smith hand function evaluation. *Am. J. occup. Ther.* **27**, 244–51.

Index